HOME INSPECTION CHECKLISTS

111 Illustrated Checklists and Worksheets You Need before Buying a Home

NORMAN BECKER, P.E.

McGraw-Hill

New York   Chicago   San Francisco   Lisbon   London   Madrid
Mexico City   Milan   New Delhi   San Juan   Seoul
Singapore   Sydney   Toronto

The **McGraw-Hill** Companies

Cataloging-in-Publication Data is on file with the Library of Congress

1 2 3 4 5 6 7 8 9 0   MAL/MAL   0 9 8 7 6 5 4 3

ISBN 0-07-142304-4

The sponsoring editor for this book was Larry S. Hager, the editing supervisor was Stephen M. Smith, and the production supervisor was Pamela A. Pelton. It was set in ACaslon Regular by Joanne Morbit of McGraw-Hill Professional's Hightstown, N.J., composition unit. The art director for the cover was Handel Low.

Printed and bound by Malloy Incorporated.

McGraw-Hill books are available at special quantity discounts to use as premiums and sales promotions, or for use in corporate training programs. For more information, please write to the Director of Special Sales, McGraw-Hill Professional, Two Penn Plaza, New York, NY 10121-2298. Or contact your local bookstore.

 This book is printed on recycled, acid-free paper containing a minimum of 50% recycled, de-inked fiber.

Contents

Introduction   v

EXTERIOR

1   Roofs   1

2   Roof-Mounted Structures and Projections   9

3   Paved Areas around the Structure   21

4   Walls, Windows, and Doors   27

5   Lot and Landscaping   37

6   Garage   47

7   Wood-Destroying Insects and Rot   55

INTERIOR

8   Attic   65

9   Interior Rooms   73

10   Basement and Crawl Space   83

ELECTROMECHANICALS

11   Electrical System   93

12   Plumbing   99

13   Heating System   109

14   Domestic Hot Water   117

15   Air-Conditioning   121

16   Swimming Pool   127

OTHER

17   Environmental Concerns   131

Glossary   135

Introduction

This book is intended for homebuyers who want to do a preliminary inspection of the house that they are considering purchasing. It is a series of inspection checklists that is also an easy-to-carry field guide that outlines the items of concern and value to a homebuyer. Since all houses are made up of many systems and components, knowing the condition of these items will allow a buyer to make a decision based on facts rather than emotion.

So that the process of evaluation may be accomplished most effectively, this book has been divided into three main sections: Exterior, Interior, and Electromechanicals. These sections are further subdivided into the following chapters covering various systems and components:

Exterior   Roofs; Roof-Mounted Structures and Projections; Paved Areas around the Structure; Walls, Windows, and Doors; Lot and Landscaping; Garage; Wood-Destroying Insects and Rot

Interior   Attic; Interior Rooms; Basement and Crawl Space

Electromechanicals   Electrical System; Plumbing; Heating System; Domestic Hot Water; Air-Conditioning; Swimming Pool

The checklists in this book have been designed to include all the different types of systems and components that may be found in a house, and all the conditions they may be found in. Therefore, there are more items in the checklists than will be found in the house being evaluated. For example, a house may be heated using steam, hot water, or warm air; and a roof may be covered with asphalt shingles, wood shingles, slate, or tile. Check off only those items that apply. In evaluating the condition of a system or component, several types of problems may exist. Check off the ones that apply. For example, asphalt shingles may be curling _____, cracked _____, torn _____, or missing _____.

The checklists are also useful when two or three houses are being considered. They will help determine which of the houses will require the least maintenance. The checklists have a provision for inspecting three houses. It is recommended that for each house a different color pencil or pen be used.

For more information on home inspection, and a detailed discussion of the checklist items, see my book *The Complete Book of Home Inspection* (3d ed., McGraw-Hill, 2002).

Norman Becker, P.E.

1
Roofs

Every roof has two basic elements: the deck and the weather-resistant covering. A proper roof inspection includes an evaluation of both. You can check the condition of the roof deck during your attic inspection. The use of binoculars is recommended for a pitched roof inspection. Try to look at all the roof slopes, especially the southerly and southwesterly exposures. Those slopes get the maximum sun, which accelerates the aging of the roof covering.

CHECKPOINTS

HOUSE

#1	#2	#3	PITCHED ROOFS

☐ ☐ ☐   Visually inspect all portions (slopes) of the roof.

☐ ☐ ☐   Are there any sagging, uneven, damaged, or patched sections? Yes _____ No _____ The cause for sagging and uneven sections can be checked during the attic inspection.

☐ ☐ ☐   Is the area directly below the roof deck ventilated? Yes _____ No _____ Look for gable vents, roof vents, or soffit and ridge vents. The lack of ventilation can cause the roof deck to delaminate.

☐ ☐ ☐   Are there any overhanging tree limbs or branches that can cause damage to the roof structure as well as the roof covering? Yes _____ No _____

☐ ☐ ☐   Is the roof in need of a cleaning to remove debris? Yes _____ No _____
A buildup of leaves, seed pods, pine needles, and twigs can impede the runoff
of rainwater, resulting in leaks.

Asphalt shingles

☐ ☐ ☐   Look for curling _____, cracked _____, torn _____, or missing _____
shingles. **Figure 1-1** If any are noted, repair or replacement is needed.

☐ ☐ ☐   Are shingles losing their stone granules? Yes _____ No _____ If so, it is a
sign of aging.

☐ ☐ ☐   Look for eroded sections in the slots between the shingle tabs. Eroded sections
are a path for leakage. Do you see any? Yes _____ No _____

☐ ☐ ☐   Pay particular attention to slopes with a southerly or southwesterly exposure.
Is there a difference between the shingles on the northerly and southerly exposures? Yes _____ No _____ The shingles on the southerly exposures normally
deteriorate more rapidly than the shingles on the northerly exposures. **Figure 1-2**

☐ ☐ ☐   Have shingles deteriorated to a point where they should be replaced?
Yes _____ No _____ **Figure 1-3**

☐ ☐ ☐   How old are the roof shingles? _____ years old.

FIGURE 1-1   Aging asphalt shingles. Note curling of the edges, with some pitting.

FIGURE 1-2 The orientation of the house can affect the projected life of the roof shingles. The deteriorated shingles on the right slope have a southerly exposure, while the shingles on the left slope have a northerly exposure.

FIGURE 1-3 Deteriorated asphalt roof shingles. Note torn, missing, and brittle shingles with a loss of the granule covering, exposing the roof mat.

☐ ☐ ☐ How many layers of shingles are there? _____ There shouldn't be more than two, although some communities allow three.

☐ ☐ ☐ If the roof has been recently reshingled, is a guarantee/warranty available?

Wood shingles, shakes

☐ ☐ ☐ Look for rotting _____, loose _____, cracked _____, chipped _____, or missing _____ sections. Any noted? Yes _____ No _____ If any are noted, repair or replacement is needed.

☐ ☐ ☐ Pay particular attention to the slopes with heavy shade.

☐ ☐ ☐ Is there a buildup of moss? Yes _____ No _____ If so, it should be removed. Moss functions like a wick; the root system provides a direct path for water entry.

Slate, asbestos-cement, clay tiles

☐ ☐ ☐ Look for missing _____, cracked _____, chipped _____, flaking _____, or loose _____ sections. If any are noted, repair or replacement is needed.

☐ ☐ ☐ Are any of the slate shingles ribbon slate? **Figure 1-4** Yes _____ No _____ Ribbon slate is an inferior-quality roof shingle.

FIGURE 1-4 Slate roof shingles. The ribbon slate is of inferior quality. Cracking often occurs along the ribbon after only 10 years.

FIGURE 1-5 Valley joint filled with a heavy layer of asphalt cement is an indication of a problem condition.

□ □ □   Are sections patched with asphalt cement? Yes _____ No _____ If so, anticipate future repairs.

□ □ □   Are the valley joints filled with asphalt cement? **Figure 1-5**
Yes _____ No _____ This is an indication of a problem condition.

□ □ □   Are there snow guards along the edge of the roof to keep the snow from sliding off? Yes _____ No _____ **Figure 1-6**

FLAT ROOFS

□ □ □   Is safe access available? Yes _____ No _____

□ □ □   Are there any cracked _____, blistered _____, eroded _____, split _____, punctured _____, or torn _____ sections? Yes _____ No _____ If so, repair is needed.

□ □ □   Look for open joints and seams. **Figure 1-7** Any noted? Yes _____ No _____ If any, they should be sealed. Open joints and seams are a cause of roof leakage.

□ □ □   Look for areas of ponding water or low points where water will accumulate.

□ □ □   Is the drainage system functional? Yes _____ No _____

□ □ □   Is the area below the roof deck adequately ventilated? Yes _____ No _____

FIGURE 1-6   Snow guards along the edge of the roof will help keep the snow from sliding off the roof.

FIGURE 1-7   Cracked and open joint in the roof parapet wall will result in leakage.

Built-up roofs (BURs)

☐ ☐ ☐ Are there areas with missing aggregate? If so, replacement of aggregate is needed.

☐ ☐ ☐ Look for patched areas _____, surface erosion _____, and alligatoring _____. **Figure 1-8**

☐ ☐ ☐ Look for wrinkled sections and blisters in the membrane.

☐ ☐ ☐ Look for open joints and seams. Any noted? Yes _____ No _____ If any, they should be sealed. Open joints and seams are a cause of roof leakage.

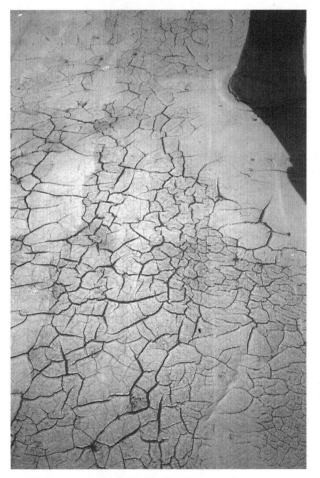

FIGURE 1-8 Roof surface with alligator-type cracks.

FIGURE 1-9   Weathered, cracked, and aging roof membrane.

☐ ☐ ☐   During the interior inspection, check the ceilings of the upper level rooms for water stains.

Roll roofing (tar paper)

☐ ☐ ☐   Are there drying, eroding, or blistered sections? These are indications of aging.
Figure 1-9

☐ ☐ ☐   Look for open joints and seams. Any noted? Yes _____ No _____ If any, they should be sealed. Open joints and seams are a cause of roof leakage.

2
Roof-Mounted Structures and Projections

When inspecting the roof, you should also inspect all roof-mounted structures and projections. Specifically look at the chimney, plumbing vent stacks, roof vents, roof hatch, skylights, TV antennas, and gutters and downspouts. These items can also be checked with the help of binoculars.

CHECKPOINTS

HOUSE
#1 #2 #3   CHIMNEYS

Masonry type (brick, stone, concrete block)

☐ ☐ ☐ Inspect for cracked _____, loose _____, chipped _____, eroding _____, or missing _____ sections of masonry. If any are noted, repair or rebuilding is needed. **Figure 2-1**

☐ ☐ ☐ Check mortar joints for cracked _____, loose _____, and deteriorating _____ sections. If any are noted, joints will need repointing.

☐ ☐ ☐ For stucco-finished chimneys, look for cracked _____, chipped _____, missing _____, or loose _____ sections of stucco.

FIGURE 2-1 Deteriorated chimney with no flue liner. Chimney should be rebuilt and flue should be lined.

☐ ☐ ☐ Is the chimney vertical or leaning? Yes _____ No _____ If it is leaning, bracing is needed.

☐ ☐ ☐ Are there open joints between the chimney and the sidewall? **Figure 2-2** Yes _____ No _____ Open joints should be sealed.

☐ ☐ ☐ If the roof is flat, does the chimney extend 3 feet above the roofline?

☐ ☐ ☐ If the roof is pitched, does the chimney extend 2 feet above the roof ridge or above any part of the roof within 10 feet measured horizontally? **Figure 2-3**

☐ ☐ ☐ If possible, check for cracked _____ or missing sections _____ of the chimney cap.

☐ ☐ ☐ If possible, check the chimney flashing for holes _____, tears _____, or loose sections _____. **Figure 2-4** Check these vulnerable areas for leakage again during your attic inspection.

☐ ☐ ☐ If possible, check to see if the chimney flue is lined.

☐ ☐ ☐ Is there a chimney top damper? **Figure 2-5** Yes _____ No _____

☐ ☐ ☐ Is there a cricket (saddle) behind the chimney? **Figure 2-6** Yes _____ No _____ A cricket is recommended if the chimney is more than 2 feet wide.

FIGURE 2-2 Uneven settlement of chimney. Note the open joint between the chimney and the sidewall.

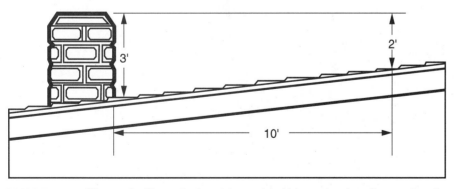

FIGURE 2-3 Chimneys should extend at least 3 feet past the highest point where they pass through the roof, and 2 feet higher than any roof part within 10 feet measured horizontally.

FIGURE 2-4 Open and loose sections of flashing caused by movement of the chimney.

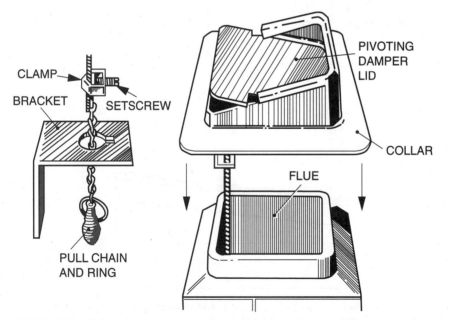

CLAMP

BRACKET

SETSCREW

PULL CHAIN
AND RING

PIVOTING
DAMPER
LID

COLLAR

FLUE

FIGURE 2-5 Chimney top damper controlled by pull chain mounted in fireplace.

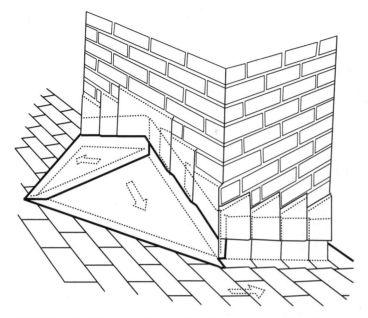

FIGURE 2-6 Cricket or saddle is used to divert water around the chimney.

Metal type (prefabricated)
_____

☐ ☐ ☐ Check for corrosion holes _____, rusting _____, or missing sections _____.

☐ ☐ ☐ Is a rain cover present? Yes _____ No _____

☐ ☐ ☐ Note condition of flashing and seal around the roof joint.

VENT STACKS

☐ ☐ ☐ Plumbing vent stacks visible? Yes _____ No _____ If not, their absence can be verified during attic inspection.

☐ ☐ ☐ Are the joints between the vent stacks and the roof sealed with asphalt cement? Yes _____ No _____ If yes, periodic application of asphalt cement should be anticipated.

☐ ☐ ☐ Note any questionable roof joints and check further during attic inspection.

☐ ☐ ☐ Do any vent stacks terminate near windows? **Figure 2-7** Yes _____ No _____ If yes and if window is open, sewer gases can be blown into the room.

☐ ☐ ☐ Do any vent stacks run up an exterior side of the house (in northern climates)? **Figure 2-8** If so, moisture in the stack can freeze over during the winter and block the opening.

FIGURE 2-7 Plumbing vent stack terminating near window. If window is open, the discharging sewer gases can be blown into the house.

FIGURE 2-8 Exterior-mounted plumbing vent stack.

Roof vents, hatches, skylights, TV antennas, dish antennas, lightning protection

☐ ☐ ☐ Check all roof joints associated with vents, hatches, and skylights.

☐ ☐ ☐ Note questionable sections and verify tightness of joints during attic inspection.

☐ ☐ ☐ On a flat roof is there a roof hatch? Yes _____ No _____

☐ ☐ ☐ Can you open it? Yes _____ No _____

☐ ☐ ☐ Are there cracks or open joints in the cover? Yes _____ No _____

☐ ☐ ☐ Check skylights for cracked _____ or broken panes _____ or defective seals (clouding) _____ between the panes. If noted, the skylight should be replaced.

☐ ☐ ☐ During interior inspection, check the ceiling area below skylights for signs of leakage.

☐ ☐ ☐ Check TV antenna connections to roof.

☐ ☐ ☐ Is antenna adequately grounded? Yes _____ No _____

☐ ☐ ☐ If there are lightning rods are they secure? Yes _____ No _____

GUTTERS AND DOWNSPOUTS

Exterior-mounted gutters

☐ ☐ ☐ Are there any missing sections of gutters? Yes _____ No _____ Missing sections will have to be replaced.

☐ ☐ ☐ Note type of material: copper _____, aluminum _____, galvanized iron _____, wood _____, vinyl _____.

☐ ☐ ☐ Are any gutter sections sagging or incorrectly pitched? If so, they should be reset.

☐ ☐ ☐ Check metal gutters for corrosion holes _____, sagging sections _____, loose support straps _____, loose support spikes _____, and leaking sections _____. **Figure 2-9**

☐ ☐ ☐ Check wood gutters for cracked sections _____ and areas of rot _____, particularly at connections and end sections. **Figure 2-10**

Built-in gutters

☐ ☐ ☐ Check for areas with rotting trim. Any noted? Yes _____ No _____

☐ ☐ ☐ Are there signs of leakage? Yes _____ No _____

☐ ☐ ☐ Note signs of seepage (stains) in soffit trim below gutters.

☐ ☐ ☐ Where possible, check condition of gutter channel.

FIGURE 2-9   Loose gutter spike. Should be reset and resecured.

FIGURE 2-10   Wood gutters. Note the cracked and rotting corner joint.

Downspouts

☐ ☐ ☐   Note type of material: copper _____, aluminum _____, galvanized iron _____.

☐ ☐ ☐   Are there any missing downspouts? Yes _____ No _____ **Figure 2-11**

☐ ☐ ☐   Check downspouts for improper joint connections. **Figure 2-12**

☐ ☐ ☐   Inspect for loose straps _____, open seams _____, and corrosion holes at elbows _____.

☐ ☐ ☐   Do downspouts have elbows at base and extensions and splash plates (or other means) to direct roof rain runoff away from the house? Yes _____ No _____ **Figure 2-13** If not, water will accumulate around the foundation and could enter the lower level of the house.

☐ ☐ ☐   If downspouts terminate in the ground, try to find out whether they are connected to dry wells or to free-flowing outlets. **Figure 2-14** Free-flowing outlets are more effective because dry wells can become clogged.

FIGURE 2-11 Water stains on the exterior siding are the result of a missing downspout.

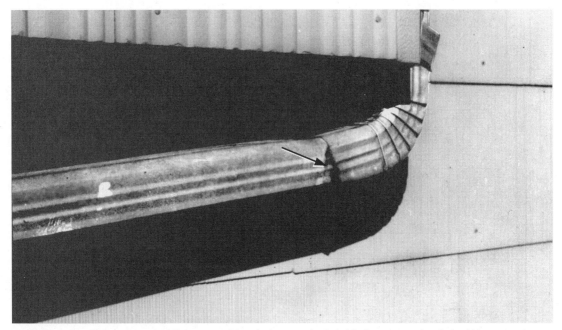

FIGURE 2-12 Incorrectly assembled downspout joint; the lower section is inside the upper section. It should be reversed; otherwise, water will leak out of the joint.

FIGURE 2-13 Splash plate directs and carries downspout effluent away from the foundation.

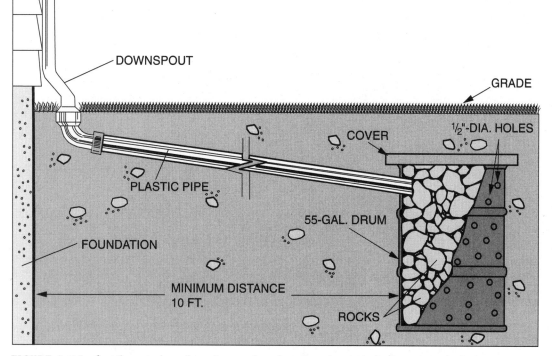

DOWNSPOUT

GRADE

COVER

½"-DIA. HOLES

PLASTIC PIPE

55-GAL. DRUM

FOUNDATION

MINIMUM DISTANCE
10 FT.

ROCKS

FIGURE 2-14 Over the years dry wells can become clogged, causing rainwater in the downspout to back up.

3
Paved Areas around the Structure

As you walk around the house inspect the paths, steps, patio, and driveway. The problems normally encountered with these items usually do not require immediate correction. Nevertheless, a tripping hazard might exist, cosmetic maintenance might be needed, or a condition might make the lower level vulnerable to water penetration. If you see problems in these areas, record them on your worksheet for early correction.

CHECKPOINTS

HOUSE

#1	#2	#3	SIDEWALKS, PATHS

☐ ☐ ☐   Inspect for cracked _____, missing _____, eroding _____, and uneven _____ sections. If any are noted, repairs will be needed.

☐ ☐ ☐   Check for areas that might present a tripping hazard, such as cracked sections, or a single step in a sidewalk or entry path. **Figure 3-1** Uneven joints can also cause a person to twist an ankle.

☐ ☐ ☐   Check the slope of all paved paths adjacent to the house for improperly pitched sections. These can result in water seepage into the lower level of the house.

FIGURE 3-1   A single step in the middle of a path is a potential hazard. The shrubs on both sides of this path call attention to the step. However, there should also be outdoor lighting in the area.

□ □ □   Is the front entry path overgrown with shrubs? Yes _____ No _____ If so, the shrubs should be pruned.

STREET-LEVEL/DRIVEWAY-LEVEL STEPS

□ □ □   Inspect for cracked _____, chipped _____, broken _____, or uneven _____ sections.

□ □ □   Are there missing handrails? Yes _____ No _____ Handrails are needed as a safety feature, especially in areas where there is freezing rain during the winter months.

ENTRY STEPS (MASONRY, WOOD)

□ □ □   Inspect for cracked _____, broken _____, loose _____ or deteriorating _____ sections.

□ □ □   Note potential tripping hazards such as variations in the height of the risers and narrow treads.

□ □ □   Check (probe) for rot in wood stringers, step treads, and handrails. Is any rot detected? Yes _____ No _____

□ □ □   Are wood stringers supported on concrete pads _____ or are they resting on the earth _____? If the latter, they are vulnerable to rot and termite activity.

FIGURE 3-2 Handrail is needed for steps, especially when they could be covered with ice.

☐ ☐ ☐ Check for handrails. Are handrails missing? Yes _____ No _____ There should be a handrail when there are more than three steps; however, two steps are preferred, especially if the steps could be covered with ice. **Figure 3-2**

☐ ☐ ☐ Inspect metal handrails for rusting _____, loose _____, and broken _____ sections.

☐ ☐ ☐ Inspect wooden handrails for cracked _____, broken _____, loose _____, and rotting _____ sections.

☐ ☐ ☐ Does entry door open onto a step or landing? Yes _____ No _____ **Figure 3-3** If so, it is considered a potential tripping hazard.

PATIO

☐ ☐ ☐ Check for cracked _____, broken _____, eroding _____, and uneven _____ areas.

FIGURE 3-3   The turnaround area of the top step is too narrow to operate the door safely.

☐ ☐ ☐   Look for uneven and settled sections adjacent to the house that can allow water to accumulate around the house foundation.

☐ ☐ ☐   Check for rot and insect damage to embedded wood sections.

DRIVEWAY

☐ ☐ ☐   Inspect for cracked _____, broken _____, eroding _____, or settled _____ areas. Note extensively cracked and deteriorated areas that will require rehabilitation. **Figure 3-4**

☐ ☐ ☐   Is the slope of the driveway level _____, raised _____, or inclined _____?

☐ ☐ ☐   For an inclined driveway, is there an adequate drain at the base? Yes _____ No _____

☐ ☐ ☐   Does the drain discharge to a dry well or to a free-flowing outlet? Yes _____ No _____

☐ ☐ ☐   Is driveway width adequate (8 feet minimum, 9 feet or more preferred)?

FIGURE 3-4 Extensively cracked and deteriorating driveway.

4
Walls, Windows, and Doors

After you walk around the house inspecting the paved areas, walk around it one more time and inspect the exterior walls, windows, trim, and doors. Also note, for further investigation, any pipe or hood projecting through a wall or basement window. These items are usually not problem conditions, but it is useful to understand their function. The hood is often covering the discharge end of an exhaust fan or clothes-dryer duct, and the pipe may be the discharge line for a sump pump or condensate line from an air-conditioning system.

CHECKPOINTS

HOUSE

#1	#2	#3	**GENERAL CONSIDERATIONS**
☐	☐	☐	Inspect exterior walls for sagging, bulging sections, and for corners that are not vertical.
☐	☐	☐	Check for window frames and door frames that are not square.
☐	☐	☐	Structural problems whose cause cannot be determined should be evaluated by a professional.
☐	☐	☐	Note wall locations that have pipe or hood projections. _____ Determine their usage: sump-pump discharge _____, condensate line _____, dryer vent _____, other _____.

☐ ☐ ☐ Check for vines growing up the exterior walls. **Figure 4-1** Any noted? Yes _____ No _____ As vines grow they can cause problems. They can lift wall and roof shingles. They can grow into mortar joints and can crush downspouts. **Figure 4-2**

FIGURE 4-1 Vines growing up and partially covering exterior shingle wall.

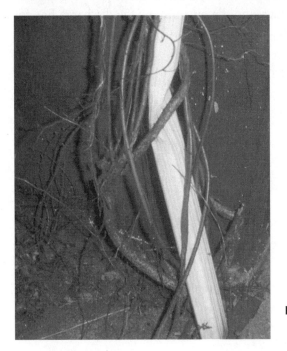

FIGURE 4-2 Downspout crushed by vines.

FIGURE 4-3   Ice layer on exterior siding caused by ice dam on roof.

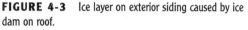

☐ ☐ ☐   If ice is noted running down an exterior wall after a snow storm, it is usually the result of an ice dam on the roof. **Figure 4-3**

EXTERIOR WALLS

Wood siding (shingles, shakes, boards, plywood panels, hardboard)

☐ ☐ ☐   Check the bottom course of the siding for sections in contact with, or in close proximity to, the ground. The bottom course should be at least 8 inches above the ground. Portions of the siding closer are vulnerable to rot and may conceal termite shelter tubes.

☐ ☐ ☐   Check wood shingles/shakes for open joints _____ and cracked _____, chipped _____, loose _____, or missing _____ sections. **Figure 4-4**

☐ ☐ ☐   Note areas of rot or discoloration. **Figure 4-5**

☐ ☐ ☐   Check for peeling and flaking paint _____ and warped _____ shingles, particularly on sidewalls with a southerly or southwesterly exposure.

FIGURE 4-4 Open shingle joints allow rainwater penetration.

FIGURE 4-5 Discoloration of wood shingles caused by rust stains from iron nails.

☐ ☐ ☐ Inspect for poor-quality shingles and shingles that have been improperly nailed.

☐ ☐ ☐ Check wood boards for open joints _____, cracked _____ and rotting _____ sections, loose _____ or missing _____ knots, and peeling _____ or blistered _____ paint.

☐ ☐ ☐ Inspect plywood panels for open joints _____ and loose _____, warped _____, cracked _____, delaminated _____, or rotting _____ sections.

☐ ☐ ☐ Check hardboard siding for cracked _____, deteriorated _____, or loose _____ sections.

Aluminum/vinyl siding

☐ ☐ ☐   Check aluminum siding for loose _____, missing _____, or dented _____ sections.

☐ ☐ ☐   Check joints for open sections and weathertightness.

☐ ☐ ☐   Does siding contain insulation backer boards? Yes _____ No _____

☐ ☐ ☐   Check aluminum siding for an electrical ground connection. (This requirement can be verified with the local building department.)

☐ ☐ ☐   Check vinyl siding for open joints and loose, cracked, or sagging sections. Any noted? Yes _____ No _____

☐ ☐ ☐   Check vinyl panels for waviness or buckling. Any noted? Yes _____ No _____ If yes, it is caused by improper nailing.

Asbestos-cement shingles/asphalt siding

☐ ☐ ☐   Check asbestos-cement shingles for loose _____, missing _____, cracked _____, chipped _____, and broken _____ sections. **Figures 4-6 and 4-7**

FIGURE 4-6 Missing and patched (mismatched) asbestos-cement shingles.

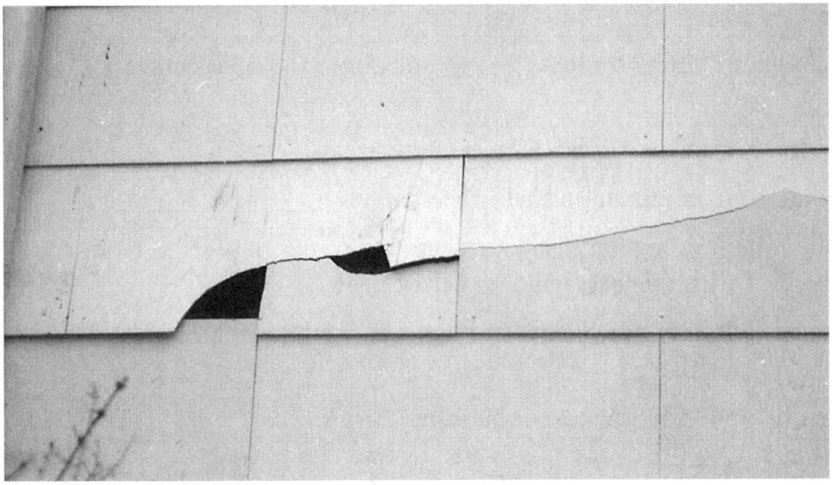

FIGURE 4-7 Cracked and chipped asbestos-cement shingles.

☐ ☐ ☐ Inspect asphalt siding for open _____ or lifting _____ joints and missing _____, loose _____, torn _____, cracked _____, chipped _____, or eroding _____ sections.

Stucco-cement-finished walls

☐ ☐ ☐ Check for bulging _____, missing _____, loose _____, cracked _____, or chipped _____ sections. Note areas in need of rehabilitation: _____ _____. **Figure 4-8**

☐ ☐ ☐ If stucco is painted, check condition._____

Synthetic stucco (EIFS)

☐ ☐ ☐ Check for cracked _____ and open joints _____ at the interface between the EIFS and windows _____, doors _____, and wall penetrations _____.

☐ ☐ ☐ Are there indications of moisture in those areas? Yes _____ No _____

Veneer and masonry walls

☐ ☐ ☐ Inspect for loose _____ or bulging _____ sections and large open cracks _____, particularly around door and window frames.

☐ ☐ ☐ Check for cracked _____, chipped _____, or missing _____ sections of brick or stone.

FIGURE 4-8 Cracked stucco wall. If wall was covered with vines, these cracks would be concealed.

☐ ☐ ☐   Inspect mortar joints for deterioration and cracked or loose sections. Any noted? Yes _____ No _____ If any are noted, repointing is needed.

☐ ☐ ☐   Check exterior surfaces on masonry walls for signs of water seepage (efflorescence).

TRIM

☐ ☐ ☐   Check trim for cracked _____, loose _____, missing _____, or rotting _____ sections.

☐ ☐ ☐   Inspect for areas of bare wood or blistered and peeling paint.

☐ ☐ ☐   Check nonwood trim for cracked _____, torn _____, missing _____, or loose _____ sections.

WINDOWS

☐ ☐ ☐   Check for cracked _____, broken _____, or missing _____ panes.

☐ ☐ ☐   Are any of the windows painted shut? Yes _____ No _____

☐ ☐ ☐   Are the panes properly secured to the sashes? Yes _____ No _____

☐ ☐ ☐   Check the condition of the window frames and sashes.

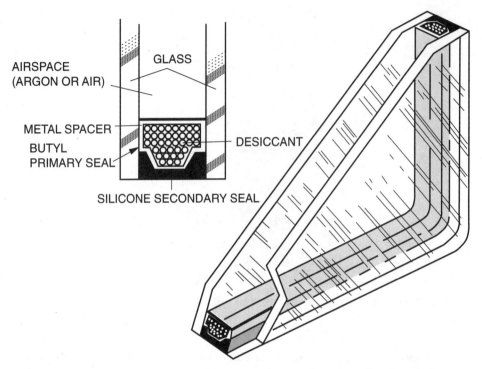

FIGURE 4-9 With an insulated glass pane, if there is a defective seal, moisture will condense between the panes.

☐ ☐ ☐   Check exterior sills for cracked and rotting sections. Any noted?
Yes _____ No _____

☐ ☐ ☐   If windows are double paned (insulated), is there clouding or condensation between the panes? Yes _____ No _____ **Figure 4-9** If there is, those panes should be replaced.

EXTERIOR DOORS

☐ ☐ ☐   Check for cracked _____, chipped _____, broken _____, or delaminating _____ sections.

☐ ☐ ☐   Is weatherstripping needed around exterior joints? Yes _____ No _____

☐ ☐ ☐   If the sliding patio doors are double paned (insulated), is there clouding or condensation between the panes? Yes _____ No _____ If there is, those panes should be replaced. **Figure 4-10**

FIGURE 4-10   Sliding insulated patio doors. Note that two of the doors have defective seals.

STORM WINDOWS, SCREENS, AND STORM DOORS

☐ ☐ ☐   Are there any missing units and/or partial installations? Yes _____ No _____

☐ ☐ ☐   Inspect storm windows for loose _____, cracked _____, broken _____, or missing _____ panes.

☐ ☐ ☐   Inspect wood units for cracked _____, broken _____, or rotting _____ sections.

☐ ☐ ☐   Inspect combination storm/screen units for loose _____, broken _____, rusting _____, or corroded _____ sections.

☐ ☐ ☐   Inspect screens for torn sections _____ and holes _____.

☐ ☐ ☐   Inspect doors for ease of operation; missing glass _____; and cracked _____, broken _____, rotting _____, or corroded _____ sections.

CAULKING

☐ ☐ ☐   Check joints for cracked, chipped, crumbly, missing, or loose areas of caulking compound.

5
Lot and Landscaping

During your walks around the outside of the house there are several items to check on in addition to the ones previously mentioned. Check the drainage around the house. The ground adjacent to the building should slope away from the house. Are there any retaining walls, decks, or fences? If so, their condition should be recorded below.

CHECKPOINTS

HOUSE

#1	#2	#3	DRAINAGE
☐	☐	☐	When approaching the house, take note of the overall topography. Is it level _____ or inclined _____?
☐	☐	☐	Are the inclined areas gently _____ or steeply _____ sloped?
☐	☐	☐	Is the house located near or at the bottom of an inclined street? Yes _____ No _____
☐	☐	☐	Note whether there is a storm drain (catch basin) nearby. _____
☐	☐	☐	Are there nearby streams or brooks? Yes _____ No _____
☐	☐	☐	Are you able to determine if the house is located in a flood plain or flood-prone area? Yes _____ No _____ **Figure 5-1**

FIGURE 5-1 Backyard flooding of a house located in a flood plain.

☐ ☐ ☐ Is the ground immediately adjacent to the house graded so that it slopes away on all sides of the structure? Yes _____ No _____ It should for proper drainage.

☐ ☐ ☐ Are there natural drainageways to direct surface water away from the house? Yes _____ No _____

☐ ☐ ☐ Are there low or level areas in the front, rear, or side yards that are vulnerable to water ponding? Yes _____ No _____

☐ ☐ ☐ Are there areas of ponded water on the lot? Yes _____ No _____

☐ ☐ ☐ Are you able to determine whether the house has footing drains?

☐ ☐ ☐ Can you locate the outlet for these and any other drainage pipes? **Figure 5-2**

RETAINING WALLS

Timber, railroad-tie, dry stone walls

☐ ☐ ☐ Inspect for missing _____, loose _____, and crumbling _____ sections of stone.

☐ ☐ ☐ Check timber and railroad-tie walls for cracked _____, loose _____, rotting _____, and heaved _____ sections.

FIGURE 5-2 Free-flowing outlet of curtain drain. Note water discharging from pipe.

☐ ☐ ☐ Are the wood-constructed walls properly anchored (tiebacks)?
Yes_____ No _____ **Figure 5-3**

Concrete, concrete block, wet stone walls

☐ ☐ ☐ Inspect for cracked _____ and heaved _____ sections.

☐ ☐ ☐ Check for loose _____, deteriorated _____, and missing _____ mortar joints.

☐ ☐ ☐ Is the wall heaving (leaning)? Yes _____ No _____ **Figure 5-4** If it is heaving, repair or possibly rebuilding will be needed.

☐ ☐ ☐ Are portions of the wall heavily covered with vines? Yes _____ No _____

☐ ☐ ☐ If yes, did you inspect these areas for cracked and heaved sections?
Yes _____ No _____

☐ ☐ ☐ Try to determine whether the area behind the retaining wall is adequately drained.

FIGURE 5-3 Timber retaining wall. Note the exposed end sections. This indicates that tiebacks are used for anchoring the wall.

FIGURE 5-4 Cracked and heaved retaining wall.

☐ ☐ ☐ Are there weep holes at the base of the wall? Yes _____ No _____ Weep holes are one method for drainage of a retaining wall.

☐ ☐ ☐ If yes, are they blocked? Yes _____ No _____

☐ ☐ ☐ Are the weep holes adequately sized and spaced?

LANDSCAPING

Lawn

☐ ☐ ☐ Inspect for holes, sunken sections, bald spots, and eroding areas.

☐ ☐ ☐ Estimate areas that will require recultivation. _____

☐ ☐ ☐ Note soft sections or ridges (possibly due to moles).

☐ ☐ ☐ Inspect terrace steps for cracked _____, loose _____, rotting _____, uneven _____, or missing _____ sections.

☐ ☐ ☐ Check steps for handrails _____, uneven treads _____, and variations in the height of the risers _____.

Shrubs

☐ ☐ ☐ Inspect shrubbery for overcrowding, dying sections, and blocked walkways or steps.

☐ ☐ ☐ Note areas in need of pruning, transplanting, or removal. _____

☐ ☐ ☐ Note areas of the house that are covered with vines. _____
Vines can damage downspouts. **Figure 5-5**

Trees

☐ ☐ ☐ Are there any dead trees and limbs, especially close to the house?
Yes _____ No _____ **Figure 5-6** If any are noted, they should be removed.

☐ ☐ ☐ Note tree limbs that are overhanging or resting on the roof.

☐ ☐ ☐ Note those trees that show evidence of rot, split sections, or insect infestation. They will require future professional evaluation.

DECKS

☐ ☐ ☐ Check and inspect the various deck components for safety rather than for appearance.

FIGURE 5-5   Downspout damaged by vines.

FIGURE 5-6   Large, dead branches are a potential safety hazard and should be removed.

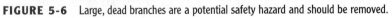

☐ ☐ ☐   Check concrete or brick piers for cracked _____, loose _____, and deteriorated _____ sections.

☐ ☐ ☐   Inspect wood columns for rot _____ and termite _____ activity.

☐ ☐ ☐   Inspect metal columns for rust deterioration.

☐ ☐ ☐   Are columns supported on concrete pads or are they in contact with the ground?

☐ ☐ ☐   Note any loose columns. _____ **Figure 5-7**

☐ ☐ ☐   Check for open and weakened joints between the deck and the house.

☐ ☐ ☐   Is the deck attached to the house with nails _____ or lag bolts _____? **Figure 5-8** (Lag bolts are preferred.)

☐ ☐ ☐   If the deck was not built at the time the house was constructed, does it have a certificate of occupancy (CO)? Yes _____ No _____

☐ ☐ ☐   Inspect deck joist supports at the portion of the deck attached to the house.

FIGURE 5-7 Inadequately supported deck. Column is loose and can easily be knocked over.

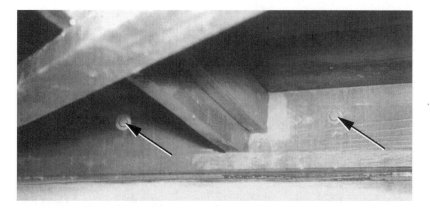

FIGURE 5-8 Lag bolts are used to secure the deck to the house. Note that joist is resting on a ledger board for support.

☐ ☐ ☐ Are deck joists supported by metal brackets _____ (preferred), or are they toe-nailed _____ into a header beam with a ledger board below the joist?

☐ ☐ ☐ Where the joists have been toenailed, check for missing ledger boards.

☐ ☐ ☐ Inspect the underside of deck (girders, joists, floor planks) for missing _____, cracked _____, and rotting _____ members.

☐ ☐ ☐ Does the deck contain diagonal bracing on the underside? Yes_____ No _____ Diagonal bracing provides additional support against twisting (racking).

☐ ☐ ☐ Inspect wood step treads, stringers, and handrails for cracked _____, loose _____, missing _____, and rotting _____ sections.

☐ ☐ ☐ Are the stringers supported on a concrete pad or are they in contact with the ground? If in contact with the ground, probe those sections with a screwdriver for rot and/or termite damage.

☐ ☐ ☐ Check top portion of deck for cracked _____, loose _____, missing _____, and rotting _____ sections of deck planks, railings, and railing posts.

FENCES

☐ ☐ ☐ Inspect wood fences for cracked _____, broken _____, loose _____, leaning _____, and missing _____ sections.

☐ ☐ ☐ Check for areas of rot and insect damage.

☐ ☐ ☐   Inspect metal fencing for loose _____, missing _____, leaning _____ and rusting _____ sections.

☐ ☐ ☐   Check gates (metal and wood) for sag _____, missing hardware _____, and cracked _____, loose _____, broken _____, and missing _____ sections.

☐ ☐ ☐   If property contains an in-ground pool, is the area around the pool adequately fenced off? Yes _____ No _____ (A fence is generally a legal requirement.)

6
Garage

The garage should be inspected after the exterior inspection has been completed. There are two basic types of garages: attached and detached. An attached garage is part of the main building. It might be located below a habitable portion of the structure or connected to the side of the building. A detached garage is a separate structure, not part of the main building. However, there might be a connecting breezeway or porch between the two structures.

CHECKPOINTS

HOUSE

#1	#2	#3	ATTACHED GARAGE

Inspecting for fire and health hazards

☐ ☐ ☐ Are the garage and basement areas combined into one open area?
Yes _____ No _____ They should be separated by a fire-rated wall.

☐ ☐ ☐ Is the interior garage door located at least one step above the garage floor?
Yes _____ No _____ Although many houses do not have a step, a step is desirable because it will prevent exhaust gases, which are heavier than air, from entering the house when the door is opened.

☐ ☐ ☐ Does this door have a tight seal and is it self-closing? Yes _____ No _____

☐ ☐ ☐   Is this door fire-resistant or does it have a sheet-metal covering on the garage side? Yes _____ No _____

☐ ☐ ☐   Is the boiler/furnace unit located in the garage? **Figure 6-1**
Yes _____ No _____ If so, the burner should be at least 18 inches above the floor level or there should be a wall at least 18 inches high around the unit.
Figure 6-2

☐ ☐ ☐   Has the boiler/furnace been placed on a raised slab? Yes _____ No _____

☐ ☐ ☐   Inspect garage ceiling and walls for exposed wood frame members. Are any noted? Yes _____ No _____ Those areas should be covered with fire-rated gypsum board. Note that the ceiling of the attached garage in Figure 6-1 should have been covered with fire-rated gypsum board.

☐ ☐ ☐   Check the ceiling area for open or missing attic access hatch.

☐ ☐ ☐   Check for return grilles in warm-air heating systems. There should **not** be any in the garage.

FIGURE 6-1 Furnace located in the garage. Note that the ceiling should have been covered with fire-rated gypsum board.

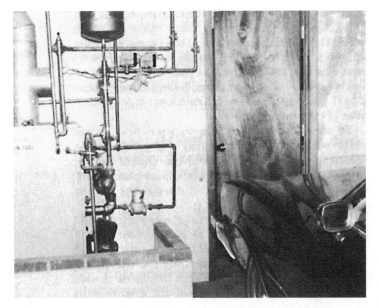

FIGURE 6-2 Heating system (oil-fired, forced hot water) boiler located in the garage. Note the low wall around the unit.

General considerations

☐ ☐ ☐ Inspect ceiling area for signs of plumbing leaks _____, water stains _____, and patched _____ sections. **Figure 6-3**

☐ ☐ ☐ If garage is unheated, are there uninsulated water pipes that are vulnerable to freezing? Yes _____ No _____

☐ ☐ ☐ Inspect floor for extensively cracked, settled, and heaved sections.

☐ ☐ ☐ Check the floor areas for evidence of water seepage and silt deposits.

☐ ☐ ☐ Does driveway incline make garage vulnerable to flooding? Yes _____ No _____

☐ ☐ ☐ Is there a drain protecting the garage entry from water penetration? Yes _____ No _____ Is it adequate? Yes _____ No _____

☐ ☐ ☐ Does garage floor contain a drain? Yes _____ No _____

☐ ☐ ☐ Inspect exterior doors and trim for cracked _____, missing _____, rotting _____, and insect-damaged _____ sections.

☐ ☐ ☐ Operate doors. Do they open and close properly? Yes _____ No _____

☐ ☐ ☐ Are there broken springs _____, missing springs _____, missing guide wheels _____, missing locks _____, and misaligned tracks _____?

FIGURE 6-3   Plumbing leak caused water-damaged ceiling in garage.

☐ ☐ ☐   Is there a restraining cable running through each spring?
Yes _____ No _____ **Figure 6-4** A restraining cable is a safety feature. It prevents the spring from whipping around if it breaks, perhaps injuring a person or damaging a car.

☐ ☐ ☐   If the overhead door is electrically controlled, does it reverse its downward travel when an upward force is exerted on the door? Yes _____ No _____ If not, the unit should be repaired or replaced.

☐ ☐ ☐   Check overhead lights, wall switches, and convenience outlets.

DETACHED GARAGE

Exterior

☐ ☐ ☐   Inspect walls/siding for bulging _____, cracked _____, loose _____, missing _____, and rotting _____ sections.

☐ ☐ ☐   Are there any broken windows? Yes _____ No _____

☐ ☐ ☐   Check roof beams for cracked _____, rotting _____, and sagging _____ members.

☐ ☐ ☐   Inspect roof shingles (as outlined in Chapter 1).

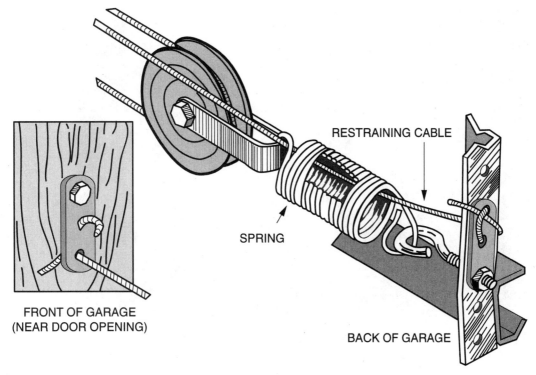

FRONT OF GARAGE
(NEAR DOOR OPENING)

RESTRAINING CABLE

SPRING

BACK OF GARAGE

FIGURE 6-4 A restraining cable through the garage door spring prevents the spring from whipping around if it breaks.

☐ ☐ ☐ Do shingles need repair or replacement? Yes _____ No _____

☐ ☐ ☐ Check type and condition of gutters and downspouts. Note their absence.

☐ ☐ ☐ Inspect and probe wood framing and trim around doors for termite damage and rot (particularly doors that are in contact with, or in close proximity to, the ground). **Figure 6-5**

Interior

☐ ☐ ☐ Check garage doors for broken _____, cracked _____, and rotting _____ sections.

☐ ☐ ☐ Inspect doors for operation, sagging sections, missing hardware, and broken glass panes.

☐ ☐ ☐ Inspect underside of roof for damaged sheathing and signs of leakage.

☐ ☐ ☐ Do the foundation/retaining walls have cracked, bowed, and heaved areas? Yes _____ No _____ If so, repairs or rehabilitation will be needed. **Figure 6-6**

FIGURE 6-5 Termite infestation and rot at base of garage door frame.

FIGURE 6-6 Heaving garage wall. Section of wall was located below grade level.

☐ ☐ ☐   Is the floor made of concrete _____ or asphalt _____, or is it a dirt floor _____?

☐ ☐ ☐   Check concrete or asphalt floor for cracked, broken, and heaved sections.

☐ ☐ ☐   Probe wood sills for insect infestation damage (particularly if these members are in contact with the ground).

☐ ☐ ☐   Are there any loose and hanging electrical wires _____, exposed junction boxes _____, extension cord wiring _____, or makeshift wiring _____?

☐ ☐ ☐   Is there a space heater? Yes _____ No _____ If so, is the heater operational? Yes _____ No _____

☐ ☐ ☐   Is the heater unit properly vented? Yes _____ No _____

☐ ☐ ☐   Are wood frame members in contact with the exhaust stack? Yes _____ No _____ There shouldn't be any contact between the wood and the exhaust stack because this would be considered a fire hazard.

7
Wood-Destroying Insects and Rot

There are many types of wood-destroying insects: subterranean and drywood termites, carpenter ants, and powder-post beetles. The one that causes the most damage to residential structures in the United States is the subterranean termite. From a distance, subterranean termites look like white or cream-colored ants. **Figure 7-1** Usually you see the shelter tubes that the termites build rather than the termites themselves. Wood will also lose its structural properties when subjected to damp conditions that are conducive to the growth of decay (rot) fungi. Check for rot and wood-destroying insects during both the exterior and interior inspections.

CHECKPOINTS

HOUSE

#1 #2 #3   **SUBTERRANEAN TERMITES**

Exterior inspection

☐   ☐   ☐   Check and probe all exterior areas of the structure that have wood in contact with, or are in close proximity to, the ground.

☐   ☐   ☐   Look for termite shelter tubes on the foundation walls. **Figure 7-2** Any found? Yes _____ No _____ If termite tubes are found, the entire house will require termite-proofing.

FIGURE 7-1 Subterranean termite activity.

FIGURE 7-2 Termite shelter tubes at base of foundation wall.

FIGURE 7-3   Termite shelter tubes on a fence post.

☐ ☐ ☐   Probe vulnerable areas such as garage-door frames, basement window sills and frames, deck posts, step stringers, and entry-door risers. Is there evidence of past or current termite activity? Yes _____ No _____

☐ ☐ ☐   Probe wood frame members adjacent to concrete-covered, earth-filled porches. Is there evidence of past or current termite activity? Yes _____ No _____

☐ ☐ ☐   Inspect crawl areas under steps and porches. Probe sills and headers. Is there evidence of past or current termite activity? Yes _____ No _____

☐ ☐ ☐   Check wood fencing, dead tree stumps, wood debris, or stored firewood in close proximity to the house for infestation and rot. Is there evidence of past or current termite activity? **Figure 7-3** Yes _____ No _____

Interior inspection

☐ ☐ ☐   Inspect for shelter tubes on foundation walls and piers. Any found? **Figure 7-4** Yes _____ No _____ No matter how small, if termite tubes are found, the entire house will require termite-proofing.

FIGURE 7-4 Termite shelter tubes on foundation wall, header, and subflooring.

☐ ☐ ☐   Pay particular attention to areas around the heating system and where the inlet water pipe and drain pipe penetrate the foundation wall or floor slab for termite shelter tubes. **Figure 7-5** Termites can always find a way into a house. **Figure 7-6**

☐ ☐ ☐   Usually in the early spring, certain termites fly out of the nest (swarm). Do you know the difference between a swarming termite and a swarming ant? **Figure 7-7**

☐ ☐ ☐   Look around basement window sills for the discarded wings of swarming termites. Any found? Yes _____ No _____

☐ ☐ ☐   Probe exposed sill plates, headers, joists, and girders for termite damage. Any damage noted? **Figure 7-8** Yes _____ No _____

☐ ☐ ☐   Inspect wood support posts for infestation and rot.

☐ ☐ ☐   If the house is built on a slab, note any soft spots in the baseboard trim.

FIGURE 7-5   Termite shelter tube on drain pipe in crawl space.

DRYWOOD TERMITES

☐ ☐ ☐   Inspect property fencing for infestation.

☐ ☐ ☐   Probe exposed wood framing throughout the house, from attic to crawl space.

CARPENTER ANTS

☐ ☐ ☐   Are there any small piles of sawdust below or around wood members? Yes _____ No _____ Sawdust is an indication of carpenter ant activity. **Figure 7-9**

☐ ☐ ☐   Are there any large black ants walking around in the rooms, particularly the kitchen? Yes _____ No _____ If yes, try to follow them as they walk to the nest.

☐ ☐ ☐   Probe the sections of wood framing, siding, and trim that show evidence of decay or past wetting.

CARPENTER BEES

☐ ☐ ☐   Carpenter bees generally do not cause structural damage, but they will cause cosmetic damage to exterior trim. **Figure 7-10**

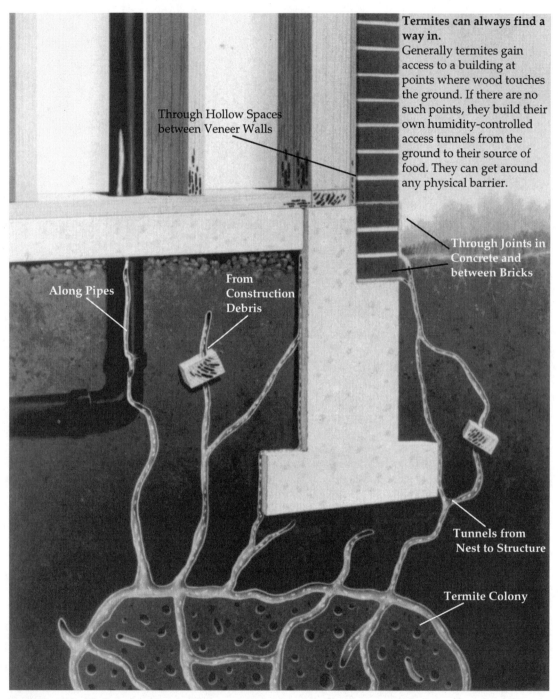

Through Hollow Spaces
between Veneer Walls

Termites can always find a way in.
Generally termites gain access to a building at points where wood touches the ground. If there are no such points, they build their own humidity-controlled access tunnels from the ground to their source of food. They can get around any physical barrier.

Through Joints in
Concrete and
between Bricks

Along Pipes

From
Construction
Debris

Tunnels from
Nest to Structure

Termite Colony

FIGURE 7-6   Termite entry points into a house.

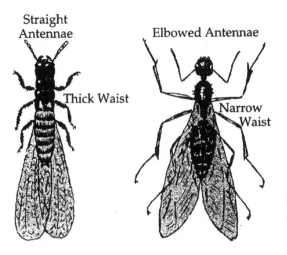

Straight Antennae

Thick Waist

Elbowed Antennae

Narrow Waist

FIGURE 7-7 The difference between swarming ant (right) and swarming termite (actual size 1/2 inch).

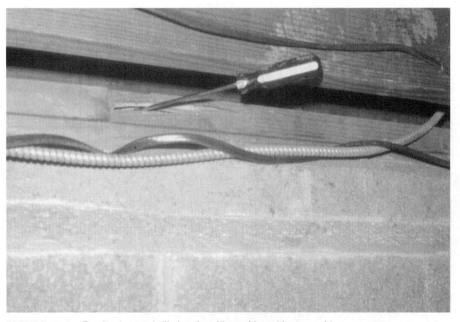

FIGURE 7-8 Termite-damaged sill plate found by probing with a screwdriver.

FIGURE 7-9 Sawdust is an indication of carpenter ant activity in window trim.

FIGURE 7-10 Carpenter bee holes in exterior trim.

POWDER-POST BEETLES

□ □ □   Inspect wood framing for clusters of small round holes. This is an indication of powder-post beetle activity. Any found? Yes _____ No _____

□ □ □   Newly formed holes are the color of a fresh saw cut and indicate an active infestation.

□ □ □   Powder-post beetle activity will cause the wood to disintegrate to the consistency of talcum powder. Probe these wood sections for deterioration. Are any sections deteriorated? Yes _____ No _____

ROT

□ □ □   Probe vulnerable areas such as wood members that are subject to periodic wetting from rain or garden sprinklers. If wood members are rotting, the probe will penetrate easily. Any rotting sections found? Yes _____ No _____

□ □ □   Inspect roof sheathing from the attic for decaying sections around chimney, roof vents, and plumbing vent stacks.

□ □ □   Check subflooring and support joists below kitchen and bathroom fixtures and around plumbing pipes for signs of rot. Any found? Yes _____ No _____

□ □ □   Probe sill plates, headers, and the ends of joists and girders for signs of rot. Any found? Yes _____ No _____

8
Attic

There are two types of attics: a full attic and a crawl attic. In a full attic a person can stand up and the walls and floor may or may not be finished (covered over). In a crawl attic the area is unfinished, and in order to move around it is necessary to crawl or stoop over. Be careful where you walk or crawl in an unfinished attic. Step only on the floor joists or your foot might break through the ceiling of the room below.

CHECKPOINTS

HOUSE
#1 #2 #3

☐ ☐ ☐   Is there access to the attic, either through a ceiling hatch or stairs?
Yes _____ No _____ If not, question the adequacy of the insulation.

INSULATION

☐ ☐ ☐   Is attic insulated? Yes _____ No _____

☐ ☐ ☐   Is the insulation a loose fill _____ or is it the batt/blanket type _____?
Table 8-1 shows typical thermal resistance numbers (R-numbers) for various types and thicknesses of insulation.

TABLE 8-1 Insulation Ratings

Insulation type	R-number					
	11	13	19	22	30	38
Batts/blankets						
Fiberglass	3½"	4"	6"	7"	9½"	12"
Rock wool	3"	4"	5½"	6"	8½"	11"
Loose-fill						
Fiberglass	5"	5½"	8½"	10"	13½"	17"
Rock wool	4"	4½"	6½"	8"	10½"	13"
Cellulose	3"	3½"	5½"	6"	8½"	11"
Vermiculite	5"	6"	9"	10"	14"	18"
Rigid board						
Polystyrene (extruded)	3"	3½"	5"	5½"	7½"	9½"
Polystyrene (bead board)	3"	3½"	5½"	6"	8½"	10½"
Urethane	2"	2"	3"	3½"	5"	6"
Fiberglass	3"	3½"	5"	5½"	7½"	9½"

☐ ☐ ☐ Is insulation properly installed? Yes _____ No _____
In a crawl attic the insulation should be in the floor. In a full attic it could be in either the walls and ceiling or the floor.

☐ ☐ ☐ Is amount of insulation adequate? Yes _____ No _____ With **Table 8-2** you can determine the amount of insulation recommended for your area.

☐ ☐ ☐ Does insulation contain a vapor barrier? Yes _____ No _____ If not, one is needed.

☐ ☐ ☐ Is vapor barrier properly installed? Yes _____ No _____ The vapor barrier should be facing the heated portion of the structure. If insulation is a loose fill, there should be a vapor barrier below it.

VENTILATION

☐ ☐ ☐ Is ventilation provided? Yes _____ No _____ Look for either gable vents _____ or a combination of ridge and soffit vents _____.

☐ ☐ ☐ If ventilation is not provided, are there any delaminating sections on the underside of the plywood roof deck? Yes _____ No _____

☐ ☐ ☐ Are vent openings blocked? Yes _____ No _____

TABLE 8-2   Recommended R-Numbers for Insulation in Ceilings, Floors, and Walls. (U.S. Dept. of Energy)

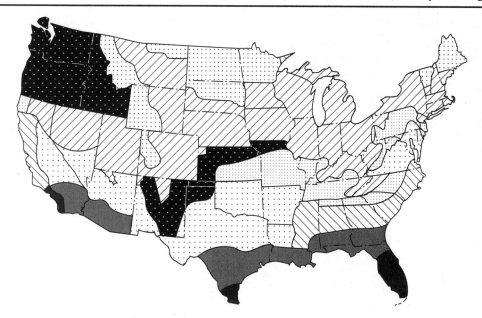

Insulation zone	Ceilings below ventilated attics		Floors over unheated crawl spaces, basements	2×4 exterior walls	2×6 walls for new construction	Crawl space walls
	Electric resistance	Gas, oil, or heat pump	*All fuel types*		*All fuel types*	
1	R-49	R-49	R-19	R-13 or R-11	R-19	R-19
2	R-49	R-38	R-19	R-13 or R-11	R-19	R-19
3	R-38	R-38	R-19	R-13 or R-11	R-19	R-19
4	R-38	R-38	R-19	R-13 or R-11	R-19	R-19
5	R-38	R-30	R-19	R-13 or R-11	R-19	R-19
6	R-38	R-30	R-19	R-13 or R-11	R-19	R-19
7	R-30	R-30		R-13 or R-11	R-19	R-19
8	R-30	R-19		R-13 or R-11	R-19	R-11

FIGURE 8-1 Water-droplet stains on catwalk in attic due to condensation on protruding roofing nails because of inadequate attic ventilation.

☐ ☐ ☐ Look on the floor for random circular water stains (**Figure 8-1**) or mildew on the roof deck. **Figure 8-2** Any noted? Yes _____ No _____ This is an indication that the attic is inadequately ventilated.

☐ ☐ ☐ Is there an attic fan _____ or roof-mounted power ventilator _____? **Figure 8-3** If so, and if weather conditions permit, check its operation.

LEAKAGE

☐ ☐ ☐ Look for water stains, which are signs of past or current roof leakage. Any noted? Yes _____ No _____ Pay particular attention to areas around the chimney and plumbing vent stacks for water stains.

☐ ☐ ☐ Is daylight visible around the joint between the vent pipe and the roof deck? Yes _____ No _____ If so, the joint should be sealed in order to prevent future leakage.

FIGURE 8-2 Mildew on underside of roofing deck.

FIGURE 8-3 Thermostatically controlled power ventilator mounted between the roof rafters.

HEATING/AIR-CONDITIONING AND EXHAUST DUCTS

☐ ☐ ☐ Are there air-conditioning and/or heating ducts in the attic?
Yes _____ No _____

☐ ☐ ☐ If so, are they insulated? Yes _____ No _____ If ducts are metal, tap them. If there is a hollow sound rather than a dull thud, there is no insulation inside.

☐ ☐ ☐ Are there any open joints in the ductwork? Yes _____ No _____ If any, they should be sealed.

☐ ☐ ☐ Are ducts from the kitchen or bathroom exhaust fans discharging into the attic? **Figure 8-4** Yes _____ No _____ They shouldn't be; they should discharge to the exterior. Sometimes the discharge end from a bathroom exhaust fan terminates in the attic floor and is covered with insulation. This negates the effectiveness of the fan.

FIGURE 8-4 Kitchen exhaust fan duct terminating in attic.

FIGURE 8-5 Cracked and broken roof rafters.

STRUCTURAL

☐ ☐ ☐ Are the rafters pulling away from the roof ridgeboard? Yes _____ No _____

☐ ☐ ☐ If visible, check the rafters and headers (framing) around skylight openings. They should be doubled up. Are they? Yes _____ No _____

☐ ☐ ☐ Are there any cracked _____, broken _____, missing _____, or sagging _____ sections of rafters or truss members? **Figure 8-5**

MISCELLANEOUS ITEMS

☐ ☐ ☐ Do any plumbing vent stacks terminate in the attic? **Figure 8-6** If so, they should be extended to above the roof.

☐ ☐ ☐ Are there open electrical junction boxes or makeshift electrical extension cord wiring in the attic? Yes _____ No _____

☐ ☐ ☐ If there is an interior type prefabricated chimney, is there fire-stopping around the joint between the chimney and the attic floor? Yes _____ No _____ An open joint is a potential fire hazard. **Figure 8-7**

FIGURE 8-6 Plumbing vent stack terminating in attic—a violation of the plumbing code. Note that in a crawl attic the proper installation of insulation is between the floor joists and not between the roof rafters.

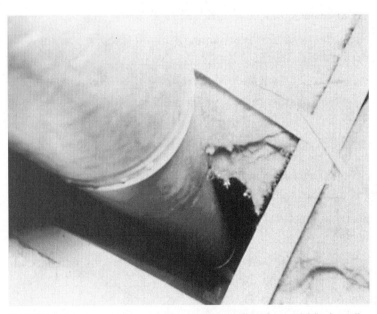

FIGURE 8-7 Open joint between chimney and attic floor. A potential fire hazard!

9
Interior Rooms

During the interior inspection, don't pass a door without opening it and looking inside. If a door is open, close it to see if it closes properly. Start your inspection of the rooms at the upper level and work your way down to the basement. As you walk from one floor to another, inspect the hallways and connecting staircase. When you walk into a room, try to look beyond the cosmetics. Look at the walls, floor, ceiling, and trim, but do not be concerned about minor cosmetic problems.

CHECKPOINTS

HOUSE
#1 #2 #3   WALLS AND CEILINGS

☐ ☐ ☐   Do not be concerned with minor cosmetic problems. Specifically check for

broken walls and ceilings _____;

loose, missing, and bulging areas of plaster or plasterboard _____;

missing, loose, and sagging sections of ceiling tile _____ (**Figure 9-1**);

sagging sections of plaster _____ (**Figure 9-2**);

truss uplift cracks (at wall-ceiling intersections) _____;

water stains on ceilings, particularly in rooms below roofs, bathrooms, and kitchens _____ (**Figure 9-3**);

disintegrating plaster, peeling and flaking paint, and water stains on walls facing the exterior (usually caused by open or exposed exterior joints) _____.

FIGURE 9-1 The ceiling had been covered with tiles, which came loose and fell down. Note that the tiles had been installed to cover a cracked, broken, and peeling ceiling.

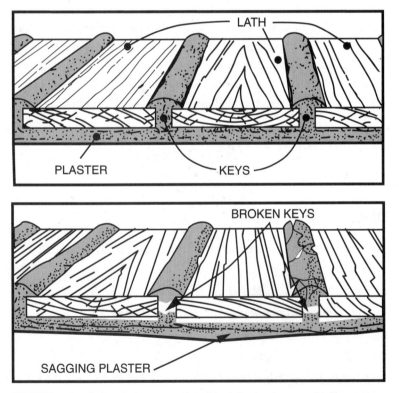

FIGURE 9-2 Ceiling plaster can sag down from lath when keys securing base coat give way.

FIGURE 9-3   Water stains on ceiling caused by leaks from radiator in room above.

☐ ☐ ☐   Note rooms that have cracked walls in which the windows and door frames are not level. Any noted? Yes _____ No _____ If yes, professional evaluation is recommended.

☐ ☐ ☐   Check trim for missing _____, loose _____, cracked _____, or broken _____ sections.

FLOORS

☐ ☐ ☐   Inspect for floors that are not level, have loose floorboards, or have sagging areas. Any noted? Yes _____ No _____

☐ ☐ ☐   Check the ceiling below a sagging floor. If the ceiling is also sagging, the condition should be evaluated by a professional.

☐ ☐ ☐   Are there any areas with large open joints between the floor and the partition walls (usually caused by excessive shrinkage)? Yes _____ No _____

☐ ☐ ☐   Inspect concrete floor slabs for cracked and settled areas. Note areas that have large open joints between the floor slab and the walls (usually caused by shrinkage of the wood framing or excessive settlement of the concrete floor slab). **Figure 9-4** Any noted? Yes _____ No _____

FIGURE 9-4 Settled concrete floor slab. Note open joint between floor and partition wall.

☐ ☐ ☐ If there are floors that have wall-to-wall carpeting, do not assume that there is a hardwood floor beneath the carpeting. Request that the owner make this representation in writing.

☐ ☐ ☐ Check raised wood floors over concrete slabs for soft, spongy, and delaminated sections. Often these conditions are a result of water seepage with associated rot in wood framing. Any noted? Yes _____ No _____

HEAT

☐ ☐ ☐ Check interior rooms for missing radiators, baseboard units, or heat registers. Any missing? Yes _____ No _____

☐ ☐ ☐ If rooms are heated by other means (radiant panels), verify this with the owner.

☐ ☐ ☐ Are radiators and heat supply registers efficiently located (preferably on an exterior wall and below a window)? Yes _____ No _____

WINDOWS

☐ ☐ ☐ Check windows for ease of operation. Do they open and close easily? Yes _____ No _____ If no, how many have problems?_____

☐ ☐ ☐ Inspect for cracked _____ and broken panes _____.

☐ ☐ ☐ Inspect for chipped _____, cracked _____, and missing _____ putty.

☐ ☐ ☐ Inspect double-hung windows for broken _____ or missing sash cords _____, loose or binding sashes _____, and missing hardware _____.

☐ ☐ ☐   Inspect steel casement and windows for cracked panes _____; rusting _____ and sprung _____ frames; and loose _____, missing _____, or inoperative _____ hardware.

☐ ☐ ☐   Do any double-pane windows have faulty seals (water droplets or cloudy areas between the glass panes)? Yes _____ No _____ If so, those panes should be replaced.

☐ ☐ ☐   Are there screens for all windows that can be opened? Yes _____ No _____

☐ ☐ ☐   Are there storm windows for all single-glazed windows? Yes _____ No _____

☐ ☐ ☐   Check casement, jalousie, and awning windows for interior storm windows and screens.

ELECTRICAL OUTLETS

☐ ☐ ☐   Inspect rooms, stairways, and hallways for electrical hazards and violations (see Chapter 11).

☐ ☐ ☐   Note all rooms in which there are insufficient outlets, and outlets that are loose or have missing cover plates.

☐ ☐ ☐   Check whether outlets are electrically "hot" and properly grounded (particularly those in the kitchen and bathrooms).

☐ ☐ ☐   Are the outlets in the kitchen and bathroom(s) protected by a GFCI (ground fault circuit interrupter)? Yes _____ No _____ If no, they should be.

FIREPLACE

☐ ☐ ☐   Inspect brick or stone firebox lining for cracked _____, chipped _____, broken _____, or deteriorating _____ sections.

☐ ☐ ☐   Check for cracked _____, loose _____, or disintegrating _____ mortar joints.

☐ ☐ ☐   Check top of firebox for an operational damper. Is it operational? Yes _____ No _____

☐ ☐ ☐   Is there a chimney-top damper? Yes _____ No _____ If so, check its operation.

☐ ☐ ☐   Inspect flue area for obstructions. Any noted? Yes _____ No _____

☐ ☐ ☐   Is the chimney flue lined? Yes _____ No _____ If no, a flue lining must be installed.

☐ ☐ ☐ Are there thick layers of soot and creosote in the chimney flue?
Yes _____ No _____ (Thick layers indicate the need for chimney cleaning.)

BEDROOMS

☐ ☐ ☐ Check that all bedrooms have at least one openable window with the following criteria:

   sill height not more than 44 inches above the floor _____;

   minimum openable area of 5.7 square feet with no dimension less than 20 inches in width _____ and 24 inches in height _____.

☐ ☐ ☐ Do all bedrooms have entry doors and closets? Yes _____ No _____

☐ ☐ ☐ Is there a smoke detector on the wall or ceiling of each bedroom?
Yes _____ No _____ If not, there should be.

☐ ☐ ☐ If a portion of the attic or basement has been converted to a bedroom, is there a certificate of occupancy for the room?

BATHROOMS

☐ ☐ ☐ Check bathrooms for adequate ventilation.

☐ ☐ ☐ If there is an exhaust fan is it operational? Yes _____ No _____ Does it have a separate on-off switch? _____ Can you determine where the fan exhaust discharges? _____

☐ ☐ ☐ Inspect tiled areas, particularly around the tub or shower, for open joints _____ and cracked _____, loose _____, and missing _____ tiles. **Figure 9-5**

☐ ☐ ☐ Note wall areas in the tub or shower that show evidence of deterioration (spongy or loose sections). Any noted? Yes _____ No _____

☐ ☐ ☐ Check shower doors for cracked panes _____ (should be safety glass) and ease of operation.

☐ ☐ ☐ Check sinks, bowl, and tub or shower for cracked _____, chipped _____, and stained _____ areas.

☐ ☐ ☐ Are the sinks and bowls properly secured? Yes _____ No _____ **Figure 9-6**

☐ ☐ ☐ Do the fixtures have individual shutoff valves? Yes _____ No _____

☐ ☐ ☐ Do any of the faucets or faucet handles leak? Yes _____ No _____ **Figure 9-7**

☐ ☐ ☐ Does the sink drain line have an S-type trap _____ or a P-type trap _____?

☐ ☐ ☐ Does the shower diverter valve in the tub leak? Yes _____ No _____

FIGURE 9-5 Missing wall tiles, leaky faucet spout, and discolored tub.

FIGURE 9-6 Sink improperly mounted and cracked.

FIGURE 9-7   Leakage around faucet handle.

☐ ☐ ☐   If there is a whirlpool bathtub, check its operation. To do this, the water level in the tub should be above the whirlpool discharge openings. Is the whirlpool operational? Yes _____ No _____

☐ ☐ ☐   Is there an access hatch for the pump motor for the whirlpool bathtub? Yes _____ No _____

Water pressure and flow

☐ ☐ ☐   Check cold-water flow by simultaneously turning on the faucets in the sink and tub or shower and flushing the bowl. Is the flow adequate? Yes _____ No _____

☐ ☐ ☐   Perform a similar check for hot-water flow. Is the flow adequate? Yes _____ No _____

☐ ☐ ☐   Check for water hammer when faucets are opened and closed rapidly.

KITCHEN

☐ ☐ ☐   Check sink for water flow and proper drainage. Are they adequate? Yes _____ No _____

FIGURE 9-8   Sink drain with an S-trap. S-traps are no longer used in plumbing installations because they can result in improper venting.

☐ ☐ ☐   If there is an island sink, is the drain vented properly or is there an S-trap?
Figure 9-8

☐ ☐ ☐   If sink contains a sprayer, check its operation (often this unit is disconnected).

☐ ☐ ☐   If sink drain contains a garbage disposal unit and house has a septic tank, you should determine the following:

   Was disposal unit added after the house was constructed? (Septic tank might be undersized.)

   When was septic tank last cleaned? (If over 3 years, the tank should be cleaned.)

☐ ☐ ☐   Inspect cabinets for missing _____, cracked _____, or loose-fitting _____ doors and drawers.

☐ ☐ ☐   Check for missing hardware on cabinet doors and drawers. Any noted?
   Yes _____ No _____

☐ ☐ ☐   Check shelving for adequate support and cracked, warped, or missing sections.

☐ ☐ ☐   Inspect counter and countertops for cracked _____, burned _____, blistered _____, or loose _____ sections.

☐ ☐ ☐   Check all appliances for operational integrity on the day of, but prior to, contract closing.

HALLWAY AND STAIRCASE

☐ ☐ ☐ Check for properly located smoke detectors in hallway areas leading to the bedrooms.

☐ ☐ ☐ Check hallways and staircases for adequate lighting. Are three-way switches located at both ends of the hallway? Yes _____ No _____

☐ ☐ ☐ Check walls, floor, ceiling, and trim as you would for interior rooms.

☐ ☐ ☐ Inspect stairways for uneven risers _____, loose treads _____, missing handrails _____, and handrails with tight finger room _____.

☐ ☐ ☐ If there is a window at the base of the stairway or landing, is the sill less than 36 inches above the floor? Yes _____ No _____ If so, a window guard should be installed.

☐ ☐ ☐ If the sill is less than 18 inches above the floor, the window should be made of tempered or safety glass.

10
Basement and Crawl Space

Problems in a basement or crawl space are often among the most costly to correct. Look specifically for signs of water penetration, structural deterioration because of rot or termite damage of the wood support members, and structural deficiencies of the foundation walls.

CHECKPOINTS

HOUSE

#1	#2	#3	FOUNDATION
☐	☐	☐	What is the foundation wall construction type: poured concrete _____, concrete block _____, brick _____, stone _____, other _____?
☐	☐	☐	Check for cracked areas of concrete _____; crumbled and flaking bricks _____; and cracked _____, loose _____, missing _____, and eroding _____ mortar joints.
☐	☐	☐	Are there any long open cracks that do not line up and have shifted sections? Yes _____ No _____ If any, the cracks are the result of differential settlement. The settlement may or may not have stabilized.
☐	☐	☐	Are there any long, open, horizontal cracks and signs of bowing in the foundation wall? Yes _____ No _____ These cracks are a serious structural problem and should be evaluated by a professional.

☐ ☐ ☐   Are sections of the structure sagging and no longer vertical?
Yes _____ No _____ If yes, consult a professional.

Wood support framing

☐ ☐ ☐   Inspect all vulnerable wood support members (sill plates, girders, joists) resting on the foundation wall for rot _____ and insect damage _____. If any is found, repair or replacement may be needed.

☐ ☐ ☐   Are there any floor joists or girders that sag or have notched sections?
Yes _____ No _____ If any is found, additional bracing of those members may be needed.

☐ ☐ ☐   Is there bridging or blocking between the floor joists? Yes _____ No _____
Figure 10-1

☐ ☐ ☐   Inspect wood columns and subflooring for cracked sections _____ and evidence of rot _____.

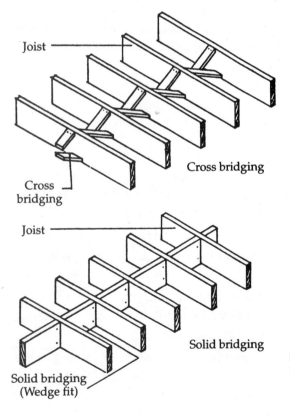

Joist

Cross bridging

Cross
bridging

Joist

Solid bridging

Solid bridging
(Wedge fit)

FIGURE 10-1   Bridging. Cross and solid bridging between floor joists.

WATER SEEPAGE

☐ ☐ ☐   Is the ground adjacent to the house pitched so that it slopes away from the structure? Yes _____ No _____ If not, it should be regraded. **Figure 10-2**

☐ ☐ ☐   Are there concrete patios and paths that have settled and are pitched toward the house? Yes _____ No _____ Settled patios or paths should be reset; otherwise, rainwater or snow melt will accumulate around the foundation and could seep into the lower level of the house.

☐ ☐ ☐   Are there basement window wells or stairwells that are vulnerable to flooding? Yes _____ No _____

☐ ☐ ☐   Do downspouts

        have extensions _____?

        discharge against the foundation _____?

        terminate in the ground _____?

☐ ☐ ☐   Is there a sump pump? Yes _____ No _____ **Figure 10-3**

☐ ☐ ☐   Is there water in the sump pit? _____

FIGURE 10-2   Finished grade of the ground adjacent to the foundation, sloped for proper drainage.

FIGURE 10-3   Sump pit in center of basement floor—a tripping hazard.

☐ ☐ ☐   If there is water in the pit, check the pump's operation. Is the sump pump operational? Yes _____ No _____

☐ ☐ ☐   Is the water in the sump pit being discharged away from the house _____ to a dry well _____ or into the house sewer line _____? Discharging to the house sewer line is illegal in most municipalities.

☐ ☐ ☐   Are there signs of water seepage? Yes _____ No _____ Check for stains at the base of exterior walls _____, a white powder (efflorescence) on the exterior walls _____ (**Figure 10-4**), rust at the base of the heating system casing _____, or rust at the base of metal support columns _____.

☐ ☐ ☐   If the structure has been waterproofed, is there a guarantee or warranty available? Yes _____ No _____

Basement rooms

☐ ☐ ☐   Check the exposed foundation wall areas for scaling _____, peeling and flaking paint _____, damp spots _____, and signs of efflorescence _____.
Figure 10-5

☐ ☐ ☐   Check construction joints, tie rod holes, and pipe openings in the foundation walls for signs of seepage.

FIGURE 10-4 Water stains and mineral deposits (efflorescence) in the corner of a foundation wall.

FIGURE 10-5 Damp spots on foundation wall.

FIGURE 10-6 Rotting and delaminating section of raised wood floor in finished basement. Condition is caused by constant wetting from water seepage into the basement.

☐ ☐ ☐ In a finished basement room, inspect wall paneling and base trim for water stains _____, warped sections _____, and rot _____. **Figure 10-6**

☐ ☐ ☐ Check underside of basement steps for water marks.

☐ ☐ ☐ Are there areas of rust at base of metal columns and sheetmetal furnace casing? Yes _____ No _____

☐ ☐ ☐ Check for musty odors and signs of mildew. These can usually be controlled with a dehumidifier. If this fails, there may be a mold problem and further investigation will be necessary.

Basement floors

☐ ☐ ☐ Check for extensively cracked and heaved floor sections (heaving is often the result of hydrostatic pressure due to a high water table).

☐ ☐ ☐ Are there areas of active seepage and puddling? Yes _____ No _____

☐ ☐ ☐ Check the joint between the foundation wall and floor slab for silt deposits. Any noted? Yes _____ No _____ These are usually an indication of past seepage through that joint. **Figure 10-7**

☐ ☐ ☐ Check for porous areas and signs of efflorescence on floor and around perimeter.

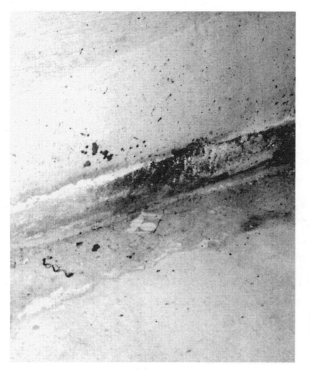

FIGURE 10-7 Signs of water seepage at joint between floor slab and foundation wall.

☐ ☐ ☐ If floor is covered with asbestos or vinyl-asbestos tile, are there swollen floor tile joints? Yes _____ No _____ Swollen joints are an indication of a water seepage problem.

☐ ☐ ☐ Is there efflorescence between joints? Yes _____ No _____

☐ ☐ ☐ If the sewer line house trap is located in the floor, is the cleanout plug secure _____ or loose _____? Some homeowners loosen or remove the plug so that water that seeps into the basement will flow down to the sewer. This is illegal in most municipalities. It is also a problem in that the sewer could back up and flood the basement.

BOILER/FURNACE ROOM

☐ ☐ ☐ Check for exposed wood frame members (wall studs, ceiling joists) that are in close proximity to the boiler or furnace. If any are noted, they should be covered with fire-rated gypsum wall board.

☐ ☐ ☐ Check for large openings between the ceiling and the chimney. If any are noted, they should be sealed as a fire-safety measure.

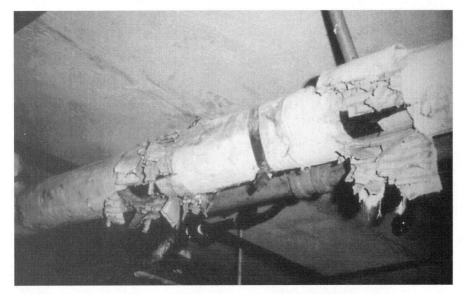

FIGURE 10-8 Deteriorating insulation on heating pipe may contain asbestos.

☐ ☐ ☐   Is the room adequately ventilated? Yes _____ No _____

☐ ☐ ☐   Is there insulation around the furnace and heating pipes that looks like it contains asbestos? Yes _____ No _____ **Figure 10-8**

CRAWL SPACE

☐ ☐ ☐   Check the foundation walls for termite shelter tubes. **Figure 10-9**

☐ ☐ ☐   Inspect foundation walls, posts, and wood support framing for signs of water seepage. Any noted? Yes _____ No _____

☐ ☐ ☐   Check the back side of the entry door to the crawl space for water stains. **Figure 10-10**

☐ ☐ ☐   Check overhead subflooring sill plates and joists for insect damage_____ and/or rot _____.

☐ ☐ ☐   Is area adequately ventilated? Yes _____ No _____

☐ ☐ ☐   Is there a dirt floor? Yes _____ No _____

☐ ☐ ☐   If there is a dirt floor, is it covered with a vapor barrier? Yes _____ No _____ If not, one should be installed. A dirt floor in the crawl space is a major source for moisture buildup in the crawl space as well as in the house.

FIGURE 10-9 Termite shelter tubes on foundation wall in crawl space.

FIGURE 10-10 Water stains on back side of entry door to a crawl area.

☐ ☐ ☐   Is area insulated? Yes _____ No _____

☐ ☐ ☐   Is insulation hanging loose or incorrectly placed? Yes _____ No _____ The insulation should be between the overhead floor joists.

☐ ☐ ☐   Are there water supply pipes in the crawl space that are vulnerable to freezing? Yes _____ No _____

☐ ☐ ☐   Are there heat supply ducts or pipes that should be insulated? Yes _____ No _____

11
Electrical System

All of the electrical wiring in a house is (or at least should be) connected to the inlet service panel box (**Figure 11-1**), which is often located in the garage or basement, although it can also be found on an interior wall within the house. These wires are called branch circuits and are protected from overloading by circuit breakers or fuses.

CHECKPOINTS

HOUSE

#1	#2	#3	**EXTERIOR**
☐	☐	☐	Is electrical service provided by underground cables _____ or overhead wires _____?
☐	☐	☐	If overhead, are the inlet service wires securely fastened to the house? Yes _____ No _____
☐	☐	☐	Count the number of the inlet service wires: 2 wires _____, 3 wires _____. **Figures 11-2 and 11-3** Note that a two-wire service is considered inadequate. It provides only 110 volts, not 220 volts, and should be upgraded.
☐	☐	☐	Inspect inlet service wires for cracked _____, missing _____, and frayed _____ sections of insulation.

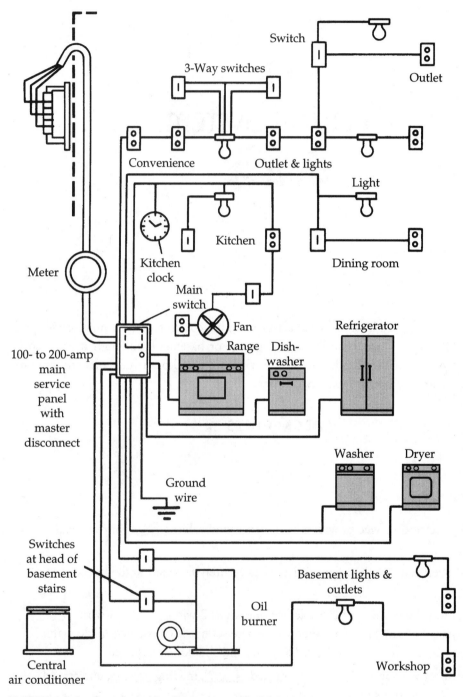

FIGURE 11-1 Generalized wiring diagram for a residential structure.

FIGURE 11-2   Two-wire inlet electrical service provides 110 volts.

FIGURE 11-3   Three-wire inlet electrical service provides both 110 and 220 volts.

☐ ☐ ☐   Note overhanging dead tree limbs or branches in contact with service wires. If there are any, they should be pruned.

☐ ☐ ☐   Is the main electrical panel box mounted on the exterior?
Yes _____ No _____ Replacement on the interior is recommended.

☐ ☐ ☐   Check outside electrical outlets for weather protection.

☐ ☐ ☐   Inspect exposed exterior wiring for proper type. It should be marked *UF-Sunlight Resistant.*

☐ ☐ ☐   Check the exterior lights and outlets for operation. Are they functional?
Yes _____ No _____

☐ ☐ ☐   Are the exterior lights and outlets protected by a ground fault interrupter (GFI)? Yes _____ No _____

☐ ☐ ☐   Note inoperative fixtures or fixtures that are missing, loose, or hanging by wires.

INTERIOR

☐ ☐ ☐   Inspect the main electrical panel box. Are the circuits protected by fuses _____ or circuit breakers _____?

☐ ☐ ☐   Was the electrical panel box made by Federal Pacific Electric?
Yes _____ No _____ There have been problems with FPE panel boxes in the past.

☐ ☐ ☐   Does system contain a main disconnect? Yes _____ No _____ It should.

☐ ☐ ☐   Note the following (do not remove the panel cover):
Loose or missing cover _____.
Missing knockout plates or fuses _____.
Are there at least two 20-amp appliance circuits? Yes _____ No _____
Is there at least one 15-amp lighting circuit for each 500 square feet of floor area?
Yes _____ No _____
Are there many spare or burned-out fuses present? Yes _____ No _____

☐ ☐ ☐   If the panel box cover is removed by an electrician or an inspector, he or she should inform you of
the amount of service: _____ amps;
circuits that are improperly protected (overfused) _____;
circuits that have aluminum wiring _____; **Figure 11-4**
evidence of water leakage or corrosion deposits _____.

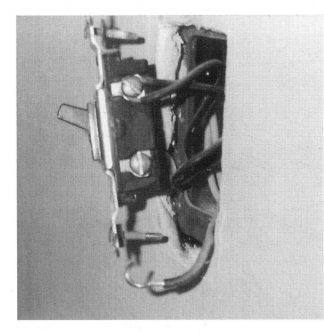

FIGURE 11-4 Aluminum wiring used for the branch circuit—a potential fire hazard.

GROUNDING

☐ ☐ ☐   Inspect electrical system for proper ground protection.

☐ ☐ ☐   If there is a municipal water supply, is the ground wire from the main panel box fastened on the street side of the water meter? Yes _____ No _____

☐ ☐ ☐   Inspect the connection for tightness of fit and corrosion.

☐ ☐ ☐   If the inlet water pipe is plastic (often in a well-pumping system), check on the exterior for a rod or pipe to which the ground wire should be clamped.

☐ ☐ ☐   Note whether the ground wire is missing _____ or has a loose _____ or corroding _____ section.

INTERIOR WIRING, OUTLETS/SWITCHES, VIOLATIONS

☐ ☐ ☐   Does house contain old or obsolete wiring (i.e., knob-and-tube-type wiring)? Yes _____ No _____ **Figure 11-5**

☐ ☐ ☐   If so, inspect for cracked and open sections of insulation.

☐ ☐ ☐   Inspect wiring in basement and attic areas for

        loose and hanging sections _____;

        extension-cord-type outlets _____;

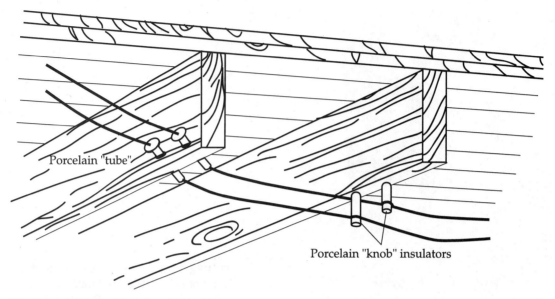

Porcelain "tube"

Porcelain "knob" insulators

FIGURE 11-5 Obsolete knob-and-tube wiring.

        open junction boxes _____;

        exposed splices _____;

        makeshift or nonprofessional alterations _____.

☐ ☐ ☐ Inspect each room in the house for electrical outlets:

        at least one outlet in the bathroom _____. (All bathroom outlets should be GFI protected.)

        at least one outlet per wall for an average-size room (10 by 12 feet) _____.

☐ ☐ ☐ Are outlets functioning (electrically hot)? Yes _____ No _____

☐ ☐ ☐ Check for loose outlets _____, switches _____, and missing cover plates _____.

☐ ☐ ☐ Are stairways and hallways adequately lit? Yes _____ No _____

☐ ☐ ☐ Are there three-way switches?

☐ ☐ ☐ Are there outlets in the hallways for night lights and cleaning equipment?

☐ ☐ ☐ Note violations such as

        open splices _____;

        fixtures hanging by wires _____;

        extension-cord wiring that passes through partitions or around door openings _____.

12
Plumbing

A basic plumbing system consists of a water supply source, distribution piping , and a waste disposal system. A major portion of the system is concealed behind the walls and below the floors. **Figure 12-1** shows the layout of a plumbing system for a typical one-family, two-story house.

CHECKPOINTS

HOUSE

#1	#2	#3	**EXTERIOR INSPECTION**
☐	☐	☐	Are there any vent stacks that terminate near windows? _____ (See Figure 2-7) run up an exterior side of the house (in northern climates)? _____ (See Figure 2-8) Both of the above locations are potential problems.
☐	☐	☐	Is the drainage system connected to a municipal sewer _____, a septic tank _____, or a cesspool _____?
☐	☐	☐	If house is connected to a septic tank, ask the homeowner where the tank and leaching field are located.
☐	☐	☐	Has the septic tank ever been cleaned? Yes _____ No _____ When? _____

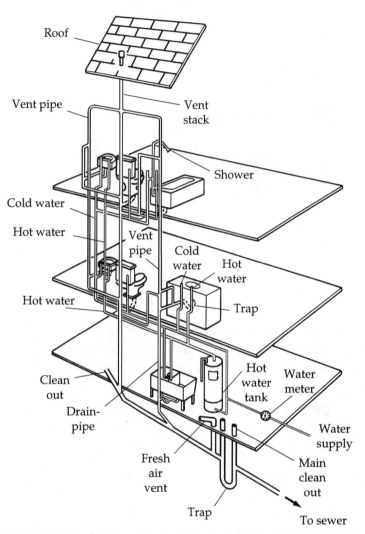

Roof

Vent pipe — Vent
stack

Shower

Cold water

Hot water — Vent
pipe   Cold
water   Hot
water

Hot water   Trap

Clean
out   Hot
water
tank   Water
meter

Drain-
pipe   Water
supply

Fresh
air
vent   Main
clean
out

Trap   To sewer

FIGURE 12-1   Plumbing system layout, showing water supply, drain, and vent pipes for a typical one-family, two-story house.

☐ ☐ ☐   Are there any wet spots or any foul odors in the area of the septic system?
Yes _____ No _____ If so, they are an indication of a problem condition.

☐ ☐ ☐   Are there any areas where liquids are oozing from the ground?
Yes _____ No _____

☐ ☐ ☐   Does property contain a lawn sprinkler system? Yes _____ No _____

☐ ☐ ☐   Is sprinkler water supply line protected by a vacuum breaker? It should be.

INTERIOR INSPECTION

Fixtures (operation and condition)

☐ ☐ ☐ Check all plumbing fixtures for operation. Are all operational?
Yes _____ No _____

☐ ☐ ☐ Note cracked _____, chipped _____, or stained _____ areas.

☐ ☐ ☐ Are any sinks or bowls loose? Yes _____ No _____

☐ ☐ ☐ Do faucets leak around handles or spouts? Yes _____ No _____

☐ ☐ ☐ Do sinks, bowls, tubs, and showers drain properly, or are they sluggish?
Yes _____ No _____

☐ ☐ ☐ Do sink and tub drain covers open and close properly? Yes _____ No _____

☐ ☐ ☐ Does sink drain have a P-trap _____ or an S-trap _____? **Figure 12-2** An
S-trap can lose its water seal and thereby allow sewer gases to seep into the
room. **Figure 12-3**

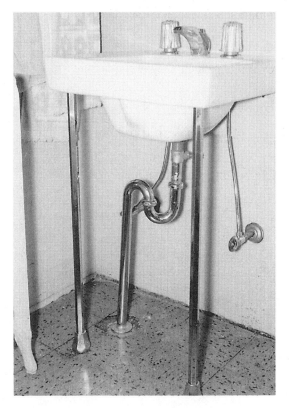

FIGURE 12-2   S-trap on sink drain.

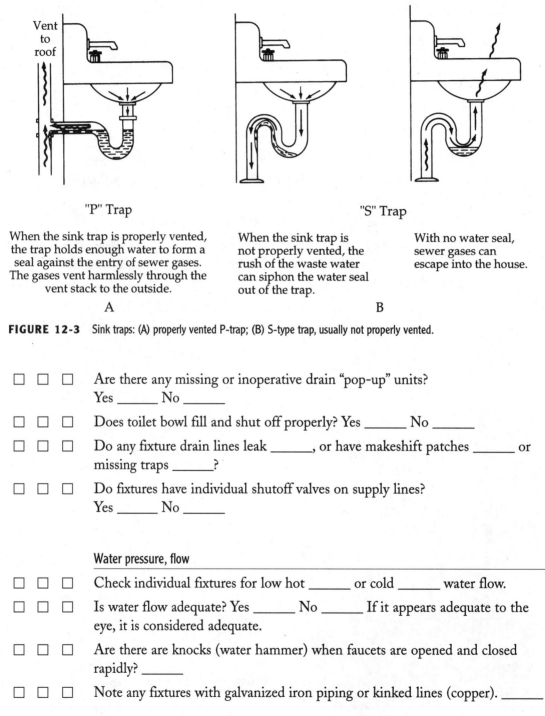

"P" Trap   "S" Trap

When the sink trap is properly vented, the trap holds enough water to form a seal against the entry of sewer gases. The gases vent harmlessly through the vent stack to the outside.

When the sink trap is not properly vented, the rush of the waste water can siphon the water seal out of the trap.

With no water seal, sewer gases can escape into the house.

A   B

FIGURE 12-3   Sink traps: (A) properly vented P-trap; (B) S-type trap, usually not properly vented.

☐ ☐ ☐   Are there any missing or inoperative drain "pop-up" units?
Yes _____ No _____

☐ ☐ ☐   Does toilet bowl fill and shut off properly? Yes _____ No _____

☐ ☐ ☐   Do any fixture drain lines leak _____, or have makeshift patches _____ or missing traps _____?

☐ ☐ ☐   Do fixtures have individual shutoff valves on supply lines?
Yes _____ No _____

Water pressure, flow

☐ ☐ ☐   Check individual fixtures for low hot _____ or cold _____ water flow.

☐ ☐ ☐   Is water flow adequate? Yes _____ No _____ If it appears adequate to the eye, it is considered adequate.

☐ ☐ ☐   Are there are knocks (water hammer) when faucets are opened and closed rapidly? _____

☐ ☐ ☐   Note any fixtures with galvanized iron piping or kinked lines (copper). _____

PIPING

Inlet service

☐ ☐ ☐ If the water is supplied by a utility company, locate the inlet pipe and the water meter if any.

☐ ☐ ☐ Is the inlet pipe made of iron _____, brass _____, copper _____, lead _____, or plastic _____? **Figure 12-4**

☐ ☐ ☐ If inlet pipe is lead, take a water sample for analysis.

☐ ☐ ☐ Is there a master shutoff valve? Yes _____ No _____ Check its operation.

Distribution piping (supply mains, fixture risers)

☐ ☐ ☐ Are these pipes copper _____ brass _____, galvanized iron _____, plastic _____, or a combination _____?

☐ ☐ ☐ Are there signs of leakage _____ or patched _____ or corroding _____ pipe sections or valves?

☐ ☐ ☐ If piping is basically brass, are there any mineral deposits along the undersides of pipes or around threaded joints? **Figures 12-5 and 12-6**
Yes _____ No _____ Mineral deposits are an indication of leakage. Affected sections should be replaced.

FIGURE 12-4 Lead inlet water pipe. Note wiped joint (bulge) near shutoff valve.

FIGURE 12-5 Brass water pipe with pinhole leaks, a condition caused by leaching zinc. Note mineral deposits on underside of pipe.

FIGURE 12-6 Brass pipe with pinhole leaks and mineral deposits. Pipe should be replaced.

☐ ☐ ☐ Pipes located in an unheated area such as a crawl space, garage, and so on may be vulnerable to freezing and should be insulated.

☐ ☐ ☐ Are any pipes improperly supported? Yes _____ No _____

☐ ☐ ☐ Are hot- and cold-water lines adequately spaced apart? Yes _____ No _____

Drainage pipes

☐ ☐ ☐ Are the pipes made out of cast iron _____, galvanized iron _____, copper _____, lead _____, or plastic _____?

☐ ☐ ☐ Look for drainage lines that are improperly pitched, or sagging sections where solid wastes can accumulate. Any noted? Yes _____ No _____ **Figure 12-7**

☐ ☐ ☐ Note any signs of leaking _____ or cracked _____ or patched _____ sections.

☐ ☐ ☐ Is there a sewage ejector pumping system below the floor slab? **Figures 12-8 and 12-9** If so, is it operational? Yes _____ No _____

WELL-PUMPING SYSTEMS

☐ ☐ ☐ Have a water sample analyzed for contamination.

☐ ☐ ☐ Is there a deep- _____ or shallow-type _____ well?

☐ ☐ ☐ Is well pump a piston _____, jet _____, or submersible _____ type?

FIGURE 12-7 Low point in drainpipe can cause solid wastes to build up and block the flow.

FIGURE 12-8   Sewage ejector pumping system below the floor slab.

☐ ☐ ☐   Do you know the design criteria (gallons/minute) of the system?
Yes _____ No _____

☐ ☐ ☐   Are installation records and recorded flow available? Yes _____ No _____

Well-pump accessory equipment

☐ ☐ ☐   Is storage tank insulated? Yes _____ No _____

☐ ☐ ☐   Are there signs of rust _____ or corroding areas _____?
Yes _____ No _____

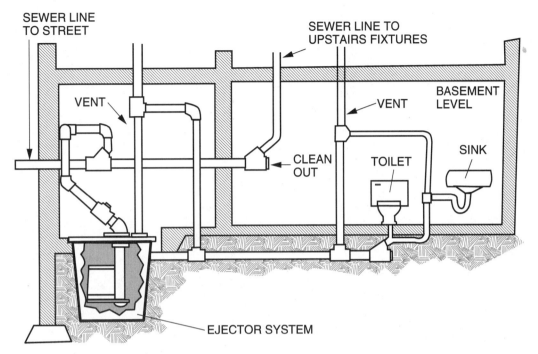

FIGURE 12-9 Sewage ejector system enables bathroom fixtures to be installed below the sewer line.

□ □ □ Does tank contain a pressure relief valve? Yes _____ No _____

□ □ □ Is pressure gauge operational? Yes _____ No _____

□ □ □ When system is active (pumping), note the pressure differential. _____

□ □ □ Does the pressure exceed 65 psi? Yes _____ No _____ If yes, the pressure is considered high and adjustment or maintenance is needed.

□ □ □ Does gauge fluctuate rapidly, or does pump rapidly cycle on and off? Yes _____ No _____ If yes in either case, it is an indication that the storage tank is waterlogged, a condition that must be corrected.

□ □ □ Does system hold its pressure when all faucets are shut and there are no interior plumbing leaks? Yes _____ No _____

13
Heating System

Central heating is supplied to most homes through a furnace or boiler. A furnace is used for producing warm-air heating, whereas a boiler is used for producing steam or hot-water heating. The heating medium for boilers and furnaces is usually oil or gas, although a furnace could also be heated by an electric resistance coil. Heating can also be supplied through a heat pump, which is basically a compressor-cycle air-conditioning system that can operate in reverse. See Chapter 15 on air-conditioning.

CHECKPOINTS

HOUSE

#1	#2	#3	**GENERAL CONSIDERATIONS**
☐	☐	☐	How old is the heating system? _____ years old
☐	☐	☐	Is the boiler or furnace an obsolete unit (i.e., a converted coal-fired boiler or an "octopus" warm-air furnace)? Yes _____ No _____ **Figures 13-1 and 13-2**
☐	☐	☐	Check each room for heat supply registers or radiators.
☐	☐	☐	Are supply registers or radiators located on or near exterior walls (preferably under the windows or next to exterior door openings)? Yes _____ No _____
☐	☐	☐	Check supply register dampers for ease of operation.

FIGURE 13-1   Old "octopus" warm-air furnace. Note the heat supply ducts at the top. Such units are inefficient and obsolete, although they can still be found in homes.

FIGURE 13-2   Old cast-iron boiler with new oil burner. Boiler was converted from a coal burner.

☐ ☐ ☐   Are there separate return grilles for each room _____ or centrally located returns for each floor _____?

☐ ☐ ☐   Check radiators for broken shutoff valves and blocked air vents (steam systems).

☐ ☐ ☐   Inspect areas below radiators for signs of leakage.

☐ ☐ ☐   Check visible portions of ductwork and piping (basement, attic, garage, etc.) for open or leaking joints and uninsulated sections. Any noted?
Yes _____ No _____

☐ ☐ ☐   Are the thermostats properly located (no drafts)? Yes _____ No _____

☐ ☐ ☐   Note whether the thermostats are manual _____ or automatic clock _____ units.

☐ ☐ ☐   Is the clock functional? Yes _____ No _____

☐ ☐ ☐   Check the operation of each heating zone and thermostat control independently.

☐ ☐ ☐   Is there an adequate air supply for the heating system boiler or furnace?
Yes _____ No _____ The unit must not be located in a tight unventilated room.

☐ ☐ ☐   If the heating unit has a direct vent, is it properly located?
Yes _____ No _____

BOILER OR FURNACE

☐ ☐ ☐   Check boiler or furnace during startup. Note any puffback with an oil burner or a licking back of flames on a gas burner.

☐ ☐ ☐   Does the boiler appear to be neglected? Yes _____ No _____

☐ ☐ ☐   Check for signs of excessive corrosion (rust), dust, flaking metal, and mineral deposits indicating past or current leakage.

☐ ☐ ☐   If system is oil fired and accessible, inspect the firebox for water seepage through joints and dripping water. If any is noted, have this condition checked by a professional.

☐ ☐ ☐   Check the smoke pipe for corrosion holes _____, open joints _____, sagging sections _____, and adequate clearance from wood _____. There should be a 2-inch clearance.

☐ ☐ ☐   If furnace or boiler is the condensing type, check for corrosion _____, signs of condensate leakage _____, and blockage in the supply and exhaust vent pipes _____ and the horizontal vent pipe sections _____ sloping back to the furnace or boiler.

☐ ☐ ☐ Check master shutoff switches for proper operation.

WARM-AIR SYSTEMS

☐ ☐ ☐ Check furnace, burner area, and sheetmetal casing for excessive corrosion, rust, and flaking metal.

☐ ☐ ☐ In a gas-fired furnace, does the burner flame appear to be unstable, like a dancing flame? Yes _____ No _____ If so, have the heat exchanger checked by a professional. This condition may indicate a cracked heat exchanger.

☐ ☐ ☐ If there is a canvas connection between furnace and main supply duct, check it for torn and open sections. This material may contain asbestos.

☐ ☐ ☐ Check fan controller during furnace operation. If fan starts after the burners shut off, controller is either faulty or out of adjustment.

☐ ☐ ☐ Check fan operation for excessive noise or vibrations. Any noted?
Yes _____ No _____

☐ ☐ ☐ If system contains a power humidifier, check for operation. Are there signs of leakage and excessive mineral deposits? Yes _____ No _____

☐ ☐ ☐ During furnace operation, check airflow and temperature at room supply registers.

☐ ☐ ☐ Is the furnace filter easily accessible for replacement or cleaning?
Yes _____ No _____

HOT-WATER SYSTEMS

☐ ☐ ☐ Check whether system has forced _____ or gravity _____ circulation. If there is no circulator pump, it is a gravity system.

☐ ☐ ☐ During operation, check the pressure gauge. The system pressure should generally be between 12 and 22 psi. Is the pressure OK? Yes _____ No _____

☐ ☐ ☐ Inspect the boiler for a relief valve, which should be boiler mounted. Do not check valve operation. If the valve is old, it may not reseat properly.

☐ ☐ ☐ Check the circulator pump(s) for proper operation (motor and pump both functioning). Note any unusual loud sounds. Any noted?
Yes _____ No _____

☐ ☐ ☐ Inspect pumps for leaking fittings or gaskets.

☐ ☐ ☐ Check each zone valve independently by activating its thermostat. Inspect valves for dripping water and deposits.

STEAM SYSTEMS

☐ ☐ ☐   Check the water level in the glass water-level gauge. The gauge should be about two-thirds full. There should not be any accumulated sediment in the gauge, and the water level should not be fluctuating (rising and dropping). Is the level OK? Yes _____ No _____

☐ ☐ ☐   Is the water supply to the boiler a manual _____ or an automatic _____ feed?

☐ ☐ ☐   Check the boiler for required safety controls: a low-water cutoff, a high-pressure limit switch, and an automatic relief valve. Are any missing? Yes _____ No _____

☐ ☐ ☐   If the low-water cutoff is an external type, check operation by flushing the unit.

☐ ☐ ☐   Inspect the boiler relief valve. Do *not* operate. If the valve is old, consider replacement.

OIL BURNERS

☐ ☐ ☐   During the exterior inspection, did the top portion of the chimney have a buildup of accumulated soot? Yes _____ No _____ If yes, the oil burner needs a tune-up.

☐ ☐ ☐   Has the oil burner been maintained or neglected? Look for a dated maintenance service card.

☐ ☐ ☐   Are there any smoke odors or soot particles in the furnace room when the oil burner is operating? Yes _____ No _____

☐ ☐ ☐   Are any abnormal noises or pulsations emitted from the oil burner during operation? Yes _____ No _____

☐ ☐ ☐   Check for cracks and open joints near the burner, particularly around the air tube, mounting plate, and front boiler doors.

☐ ☐ ☐   Check for open joints between the smoke pipe and the chimney. If any are noted, they should be sealed.

☐ ☐ ☐   Check exhaust stack for an operational draft control. Does the damper swing freely? It should. Yes _____ No _____ **Figure 13-3**

☐ ☐ ☐   If the fire chamber is accessible, check the refractory lining for cracked and deteriorating sections.

FIGURE 13-3 Draft regulator for an oil-fired boiler.

☐ ☐ ☐ Check oil burner feed lines (copper tubing). Are there kinked or exposed sections, which are vulnerable to damage? Yes _____ No _____ **Figure 13-4**

☐ ☐ ☐ Run your hand along the underside of the interior oil storage tank. Are there any signs of leakage? Yes _____ No _____

GAS BURNERS

☐ ☐ ☐ Check the gas burners for corrosion dust and clogged gas ports. If any are noted, the burners need a cleaning.

☐ ☐ ☐ Check flame pattern for safe and efficient operation. The flame should not be mostly yellow. If it is, have the system professionally cleaned and tuned up.

☐ ☐ ☐ Check draft diverter on the exhaust stack for backflow of exhaust gases.

HEAT PUMPS

☐ ☐ ☐ Check the operation of the system in the heating mode when the temperature is below 65°F.

☐ ☐ ☐ Check the outdoor evaporator coil during startup. (This unit is the compressor when the heat pump is in the air-conditioning cycle.) Listen for any unusual sounds, such as straining, groaning, or squealing.

FIGURE 13-4   Oil-fired forced-hot-water boiler with a copper tube oil-feed line that can be easily damaged.

☐ ☐ ☐   Does unit operate smoothly without shortcycling (repeated startup and shut-down)? Yes _____ No _____

☐ ☐ ☐   Is there an ice buildup on the evaporator coil? Yes _____ No _____

14
Domestic Hot Water

Most hot water for bathing and washing (domestic hot water) is produced through a separate tank-type water heater. It can also be produced through the heating-system boiler (steam or hot water), by a tankless coil, or through an indirect-fired storage water heater.

CHECKPOINTS

HOUSE
#1 #2 #3

☐ ☐ ☐ Is hot water supplied by a separate tank-type water heater _____, by a tankless coil _____, or by an indirect-fired water heater _____?

☐ ☐ ☐ If there is a tank-type heater, record the capacity (gallons) _____ and the recovery rate (gallons per hour) _____. The data are usually on the name plate.

☐ ☐ ☐ Are exhaust gases spilling out of the draft diverter? Yes _____ No _____

☐ ☐ ☐ Is a tank-type heater adequate for the house? (See following table.)

Full bathrooms	Sum of tank capacity plus recovery rate
1	70
2	90

Full bathrooms	Sum of tank capacity plus recovery rate
3	105
4	115

☐ ☐ ☐ Is hot-water flow adequate when two hot-water fixtures are turned on? Yes _____ No _____

☐ ☐ ☐ Are hot and cold supply lines to the water heater properly installed or are they reversed? Yes _____ No _____

☐ ☐ ☐ Does water heater contain a temperature/pressure relief valve? Yes _____ No _____ If not, one should be installed.

☐ ☐ ☐ Is there a discharge pipe connected to the relief valve? Yes _____ No _____ If not, one should be installed. The pipe diameter should be the same size as that of the relief valve. Anything less will negate the effectiveness of the relief valve. **Figure 14-1** In addition it should be installed so that it is vertical in a downward direction. Otherwise a backpressure could build up, which can also negate the effectiveness of the relief valve.

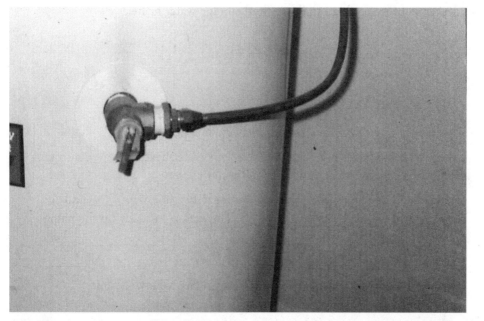

FIGURE 14-1 Improper installation of relief valve on water heater. The discharge pipe diameter should be the same size as that of the valve opening. Also, the pipe should be vertical in a downward direction.

☐ ☐ ☐ Is the relief valve dripping? Yes _____ No _____ If so, the relief valve may be defective or it may be because of thermal expansion, in which case you will need an expansion tank for the heater. **Figure 14-2**

☐ ☐ ☐ Look for signs of corrosion or past leakage:

tank type (gas-fired unit): corrosion dust and flaking metal in burner area _____

tankless coil: rust and deposits around fittings and gasket _____

☐ ☐ ☐ Is there an "instant" hot-water return line? Yes _____ No _____ **Figure 14-3**

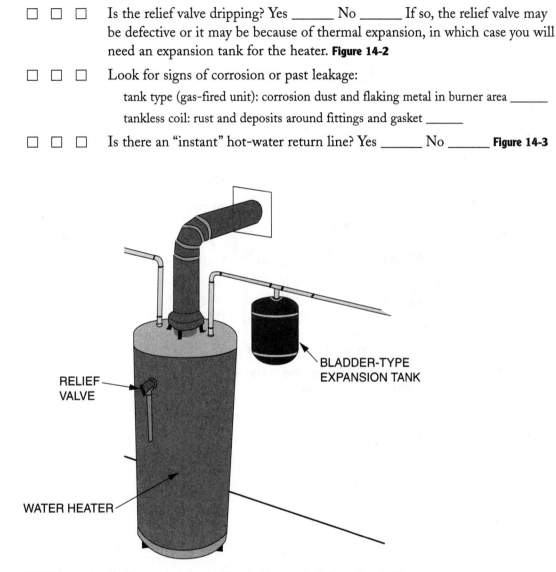

FIGURE 14-2 Bladder-type expansion tank installed to prevent relief valve from leaking due to thermal expansion of the water in the domestic water heater tank.

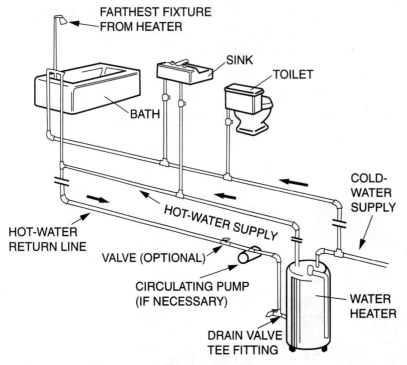

FARTHEST FIXTURE
FROM HEATER

SINK

TOILET

BATH

COLD-
WATER
SUPPLY

HOT-WATER SUPPLY

HOT-WATER
RETURN LINE

VALVE (OPTIONAL)

CIRCULATING PUMP
(IF NECESSARY)

WATER
HEATER

DRAIN VALVE
TEE FITTING

FIGURE 14-3 Hot-water return line supplies "instant" hot water to a fixture. This results from the constant recirculation of the water back to the water heater.

15
Air-Conditioning

The basic components of an air-conditioning system are the compressor, condenser, expansion device, and evaporator (cooling coil). **Figure 15-1** The compressor and condenser are located within the same metal casing, which is normally located outside the house. For a home heated with either hot water or steam, the evaporator coil along with a fan will generally be located in the attic or the basement. However, for a home heated with warm air, the evaporator coil will usually be located within the furnace casing.

CHECKPOINTS

HOUSE

#1	#2	#3	**GENERAL CONSIDERATIONS**
☐	☐	☐	How old is the air-conditioning system? _____ years old
☐	☐	☐	When was the unit last serviced? _____
☐	☐	☐	Do not turn system on if the outside air temperature is below 60° F.

COMPRESSOR-CONDENSER

☐	☐	☐	Check compressor during startup. Do you hear any unusual sounds, such as straining, groaning, or squealing. Yes _____ No _____

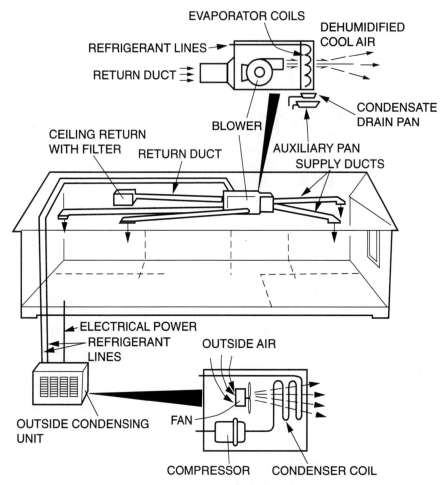

FIGURE 15-1 The basic components of an air-conditioning system in a one-story home.

☐ ☐ ☐ Does compressor operate smoothly without short cycling (repeated startup and shutdown)? Yes _____ No _____

☐ ☐ ☐ Is condenser fan operating properly?

☐ ☐ ☐ Does compressor appear to be functioning properly? (After 15 minutes, warm air should be discharging from the unit.) Yes _____ No _____

☐ ☐ ☐ Are there indications that the system is low in refrigerant (such as frosting on low-pressure refrigerant line)? Yes _____ No _____

☐ ☐ ☐ Is compressor-condenser casing located properly for maximum effectiveness (minimum sun exposure and unrestricted airflow)? Yes _____ No _____

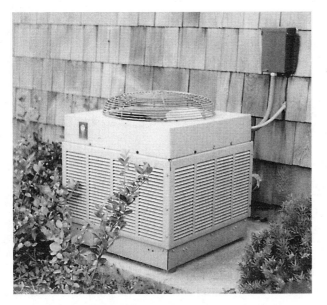

FIGURE 15-2 Typical compressor-condenser for an air-conditioning system. Airflow through the unit must be unobstructed.

☐ ☐ ☐   Is unit in need of a cleaning (clogged with leaves, twigs, dust, etc.)?

☐ ☐ ☐   Check that unit is level and adequately supported by a concrete pad or blocks.
Figure 15-2

☐ ☐ ☐   For safety and maintenance, is there a main electrical disconnect for the compressor located near the unit? Yes _____ No _____

EVAPORATOR (COOLING COIL)

☐ ☐ ☐   During operation, if possible, check cooling coil for frosting (ice buildup), usually the result of an insufficient airflow or lack of refrigerant.

Furnace-mounted evaporators (installed in the furnace plenum)

☐ ☐ ☐   Check for signs of leakage, mineral deposits, and areas of rust and corrosion.

☐ ☐ ☐   Note method of condensate discharge. Is it

to a nearby sink? _____

to the exterior? _____

to a floor drain? _____

to a hole in the floor slab (less desirable)? _____

to a small reservoir lift pump? _____

☐ ☐ ☐   If condensate discharges into a reservoir lift pump (small rectangular box), check operation of pump.

☐ ☐ ☐   Check blower motor for unusual noises and vibrations.

Blower coil (housed in a separate casing and most often located in the attic)

☐ ☐ ☐   Is unit vibration mounted? Yes _____ No _____

☐ ☐ ☐   If access is available, inspect evaporator coil for frost buildup.

☐ ☐ ☐   Check for a condensate drain line.

☐ ☐ ☐   Does this drain line discharge the condensate to a nearby roof gutter _____ or directly into a plumbing vent stack _____? **Figure 15-3** (The latter type of connection is usually not permitted and should be verified with the local building department.)

☐ ☐ ☐   Check blower coil casing for an auxiliary drain pan below the unit. The auxiliary pan should contain an independent drain line that is not connected to the main drain line.

☐ ☐ ☐   If blower coil is located in attic, is the attic adequately ventilated? Yes _____ No _____

FIGURE 15-3   Condensate drain line from air-conditioning unit in attic terminating in plumbing vent stack. In most communities this type of termination does not comply with the plumbing code.

Ducts or registers

☐ ☐ ☐ Check airflow and temperature after 15 minutes of operation.

☐ ☐ ☐ Do all rooms have air-conditioning supply registers? Yes _____ No _____

☐ ☐ ☐ If rooms do not have individual return grilles, check for a large central return grille (often located in the hall).

☐ ☐ ☐ If there is a central return, are doors to the rooms undercut to permit proper air circulation? Yes _____ No _____

☐ ☐ ☐ Check exposed ductwork in the attic and/or crawl spaces for open joints _____, signs of air leakage _____, and uninsulated ducts _____.

Heat pump

☐ ☐ ☐ If the temperature is above 65° F check the air-conditioning mode of the heat pump. When it's below that temperature, check the heating mode.

☐ ☐ ☐ When heat pump is in the air-conditioning mode, use the air-conditioning checklist above.

16
Swimming Pool

Swimming pools can be installed either in the ground or aboveground. They come in various sizes and shapes. The shape of an in-ground pool can be freeform and therefore the shapes are limitless. On the other hand, the shape of an aboveground pool is basically limited to circular, rectangular and oval. The shell of an in-ground pool will be concrete, vinyl lined with sidewall supports, or preformed fiberglass.

CHECKPOINTS

HOUSE

#1	#2	#3	**GENERAL CONSIDERATIONS**
☐	☐	☐	Has a Certificate of Occupancy been issued for the pool? Yes _____ No _____
☐	☐	☐	Is there a fence that encloses the swimming pool? Yes _____ No _____ There should be one as a safety precaution. A fence is generally a legal requirement.
☐	☐	☐	Is the fence in need of maintenance? Yes _____ No _____
☐	☐	☐	Are fence gates self-closing and self-latching? Yes _____ No _____ If not, they should be upgraded or replaced.

DECK

☐ ☐ ☐ Are there any cracked _____, chipped _____, or settled _____ sections? Yes _____ No _____

☐ ☐ ☐ Look for uneven joints between sections. Uneven joints are a tripping hazard.

☐ ☐ ☐ Cracked or open joints should be sealed. Any noted? Yes _____ No _____

☐ ☐ ☐ Are grab rails, ladder, and slide adequately anchored to deck? Yes _____ No _____

☐ ☐ ☐ Is diving board cracked or warped? _____

VINYL-LINED POOL

☐ ☐ ☐ Is lining stained, discolored, or torn? Yes _____ No _____

☐ ☐ ☐ Has lining pulled out of edge retainer? Yes _____ No _____

☐ ☐ ☐ Is bottom of liner stained?

CONCRETE POOL

☐ ☐ ☐ Are there cracked, chipped, loose, or missing tiles? Yes _____ No _____

☐ ☐ ☐ Does the plaster finish have flaking _____, chipped _____, or discolored _____ sections?

☐ ☐ ☐ Are any cracks visible in the sidewalls? Yes _____ No _____

☐ ☐ ☐ Is the painted surface flaking or faded? Yes _____ No _____

☐ ☐ ☐ Check the skimmer weir and strainer basket. **Figure 16-1**

POOL EQUIPMENT

☐ ☐ ☐ Is the pool water turbid? Yes _____ No _____

☐ ☐ ☐ Are tiny bubbles discharging into pool? Yes _____ No _____ There shouldn't be any.

☐ ☐ ☐ Is the pump noisy? Yes _____ No _____

☐ ☐ ☐ Is the pump very hot to the touch? Yes _____ No _____ If so, this may indicate that the pump is running dry, which could damage it.

☐ ☐ ☐ Check for water dripping around the pump. Any noted? Yes _____ No _____

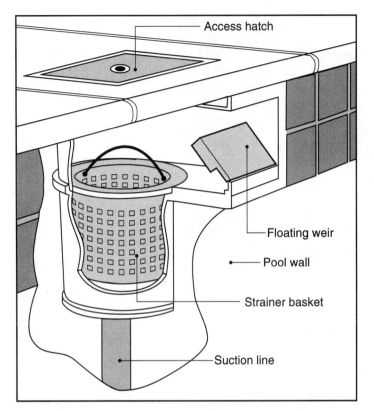

Access hatch

Floating weir

Pool wall

Strainer basket

Suction line

FIGURE 16-1 Cutaway view of a typical surface skimmer for a concrete pool.

☐ ☐ ☐ Is the pressure gauge on the filter operative?

☐ ☐ ☐ Does heater cycle on and off? Yes _____ No _____ If yes, this could indicate low water flow, a condition that should be corrected.

☐ ☐ ☐ Look for scale buildup with rust flakes and dust around heater.

☐ ☐ ☐ Is water dripping under or around the heater? Yes _____ No _____

☐ ☐ ☐ Does pool cover have torn or damaged sections? Yes _____ No _____
Figure 16-2

☐ ☐ ☐ Check the operation of underwater pool lights.

☐ ☐ ☐ Are electrical outlets GFI protected? Yes _____ No _____

FIGURE 16-2 Pool cover keeps the debris out when the pool is not in use for an extended period of time.

17
Environmental Concerns

In addition to the physical inspection, as outlined in the previous chapters, the affect of the following environmental problems should be of concern, and should enter into the decision on whether or not to purchase the house: radon, asbestos, quality of drinking water, lead paint, leaky oil tanks, electromagnetic fields, and mold.

HOUSE
#1 #2 #3   RADON

Radon concentration is measured in picocuries per liter (pCi/L). The EPA action level is 4.0 pCi/L. If the radon concentration in the lowest lived-in level is greater than 4.0 pCi/L the EPA recommends fixing the home.

☐ ☐ ☐   Has the house been tested for radon? Yes _____ No _____

☐ ☐ ☐   Is the radon concentration greater than 4.0 pCi/L? Yes _____ No _____

☐ ☐ ☐   If yes, has a radon mitigation system been installed? Yes _____ No _____

ASBESTOS

Asbestos-containing material in a house does not necessarily pose a health risk. Asbestos materials become hazardous only when, due to damage, disturbance, or deterioration over time, they release fibers into the air. Many houses have old boilers and heating pipes that are insulated with asbestos. The insulation on the old boilers looks like a white plaster coating over the boiler shell. The insulation on the pipes when viewed from the end looks like corrugated cardboard.

☐ ☐ ☐   Is the boiler shell covered with a white plaster coating? Yes _____ No _____ (See Figure 13-2)

☐ ☐ ☐   If yes, and if the boiler is to be replaced, the insulation must be removed by a qualified asbestos removal contractor.

☐ ☐ ☐   Is the insulation on heating pipes deteriorating? Yes _____ No _____ (See Figure 10-8)

DRINKING WATER

Water supplied by public water companies is usually safe to drink and does not pose a health risk. However, many homes have domestic water supplied by a private well. Private wells are vulnerable to contamination by harmful bacteria from faulty septic tanks, by leakage from buried oil storage tanks, by pesticides and fertilizers, or by chemicals from toxic spills that occurred years before. Well water should be analyzed once a year by a state-licensed testing laboratory to ensure that it is potable.

☐ ☐ ☐   Is the water supplied by a public water company _____ or a private well _____?

☐ ☐ ☐   If supplied by a well, has the water been tested for potability? Yes _____ No _____

LEAD PAINT

Lead poisoning from paint chips and lead dust particles has been a problem in older homes. According to the EPA about two-thirds of the homes built before 1940 and one-third to one-half of the homes built between 1940 and 1960 contain heavily leaded paint. A small percentage of the homes built between 1960 and 1980 had surfaces coated with lead-based paint. Houses built after 1978 should be relatively free of lead paint. The only way to tell whether the paint in a home contains lead is to have samples from different areas tested by a qualified laboratory.

☐ ☐ ☐   When was the house built? _____

☐ ☐ ☐ If the house is an older house, has the paint ever been tested?
Yes _____ No _____

LEAKY OIL TANKS

Many homeowners who have a buried oil tank for their heating system do not know that they are responsible if the tank leaks and contaminates the surrounding soil. The cost for cleanup can be in the thousands of dollars, and if the aquifer is contaminated the cost could be in the tens of thousands. Most of the older buried tanks are steel and have a projected life of about 20 years, although many units last longer. If the house that you're considering has a buried tank of that age or older, either ask the seller for certification on the integrity of the tank, or have the tank tested for leaks.

☐ ☐ ☐ Is the house heating system oil fired? Yes _____ No _____

☐ ☐ ☐ Is the oil storage tank buried _____ or is it exposed _____?

☐ ☐ ☐ If buried, is it 20 years old or older? Yes _____ No _____

☐ ☐ ☐ If yes, does the seller have a certification on the integrity of the tank?
Yes _____ No _____

ELECTROMAGNETIC FIELDS

In 1979 an epidemiological study concluded that children exposed to an electromagnetic field with a strength of 2–3 milligauss were twice as likely to develop leukemia as children living in a house without such exposure. High-voltage transmission lines near a house cause electromagnetic fields of concern, as do local power distribution lines. However, the scientific community did not universally accept the conclusions of the study because some researchers felt it was flawed. As of this writing there is no government regulation or scientific recommendation answering the question as to whether or not there is a health problem. When purchasing a house, prudent avoidance should be considered. Until the cloud of uncertainty is removed, this problem can affect the resale value of the house.

☐ ☐ ☐ Is there a high-voltage transmission line near the house?
Yes _____ No _____

MOLD

The eyes and the nose are the best tools when inspecting for mold. Mold spores can appear as black, brown, blue, orange, or white specks. They thrive on materials such as cotton, wool, paper, leather, and wood, and will grow anywhere in a house where there is moisture. The moisture can be from a water leak or even from the humidity in the air. Large mold infestations can usually be seen or smelled. In most cases they can be removed by a thorough cleaning with bleach and water. However, if there is an extensive amount of mold it should be cleaned by an experienced professional. Exposure to mold spores in a house does not always present a health problem. If there is a mold condition in a house, a person who is allergic or otherwise sensitive to mold will know it. He or she may experience symptoms such as nasal stuffiness, eye irritation, or wheezing. It is not necessary to know what type of mold exists in a house. Regardless of the type of mold, if a mold buildup is detected, it should be removed.

☐ ☐ ☐   Check walls and ceilings for mold buildup. Any found? Yes _____ No _____

☐ ☐ ☐   Is there a mold odor in any of the rooms? Yes _____ No _____
    If yes, further investigation is necessary.

☐ ☐ ☐   Pay particular attention to the basement, which is vulnerable to water seepage. Mold growth is almost always due to excessive moisture.

Glossary

Alligatoring Extensive surface cracking in a pattern that resembles the hide of an alligator.

Backflow Any flow in a direction opposite to the natural or intended direction of flow.

Cricket A small saddle-shaped projection on a sloping roof; used to divert water around an obstacle such as a chimney. Also called *saddle*.

Dry well A covered pit, either with open-jointed lining or filled with coarse aggregate, through which drainage from downspouts or foundation footing drains may seep into the surrounding soil.

Eave The lower edge of a roof that projects beyond the building wall.

Efflorescence A white powdery substance appearing on masonry wall surfaces. It is composed of soluble salts that have been brought to the surface by water or moisture movement.

Fascia (or facia) A horizontal board that is nailed vertically to the ends of roof rafters; sometimes supports a gutter.

Flashing Sheetmetal or other thin, impervious material used around roof and wall junctions to protect the joints from water penetration.

Flue A passageway in a chimney for conveying smoke, gases, or fumes to the outside air.

Gable roof A double-sloped roof from the ridge to the eaves; the end section appears as an inverted V.

Girder The main structural support beam in a wood-framed floor. The girder supports one end of each joist.

Grade The ground level existing at the outside walls of a building or elsewhere on a building site.

Header   A framing member that crosses and supports the ends of joists.

Joist   One of a series of parallel beams used to support floor and ceiling loads, and supported in turn by larger beams (girders) or bearing walls.

Pier   A masonry column, usually rectangular in horizontal cross section, used to support other structural members.

Rafter   One of a series of inclined structural roof members spanning from an exterior wall to a center ridge beam or ridgeboard.

Resilient tile   A manufactured interior floor covering that is resilient, such as vinyl or vinyl-asbestos tile.

Ridge beam   The beam or board placed on edge at the ridge (top) of the roof, into which the upper ends of the rafters are fastened.

Riser   (1) The vertical height of a stair step. (2) The vertical boards that close the space between the treads of a stairway.

Sheathing   The structural covering, usually wood boards or plywood, over a building's exterior studs or rafters.

Shelter tube   Mud-type tube (tunnel) built by termites as a passageway between the ground and the source of food (wood).

Sill plat   The lowest member of the house framing resting on top of the foundation wall. Also called *mud sill*.

Smoke pipe   A duct for carrying exhaust gases from the furnace or boiler to the chimney flue.

Soffit   The visible underside of a roof overhang or eave.

Stringer (step)   One of the enclosed sides of a stair supporting the treads and risers.

Stud   One of a series of slender wood or metal vertical structural members placed as supporting elements in walls and partitions.

Subfloor   Boards or plywood laid on joists over which a finished floor is to be laid.

Toenailing   Driving a nail at an angle into the corner of one wood-frame member in order for the nail to penetrate into a second member.

Tread   The horizontal board in a stairway on which the foot is placed.

Vacuum breaker   A device that prevents a vacuum in a water-supply system from causing backflow.

Vapor barrier   A moisture impervious layer or coating that prevents the passage of moisture or vapor into a material or structure.

Weep hole   A small opening at the bottom of a retaining wall or the lower section of a masonry veneer facing on a wood-frame exterior wall, which permits water to drain.

About the Author

A nationally known authority on the subject of home inspection, Norman Becker, P.E., has more than 31 years' experience in home and building inspection. He is one of the founders of the American Society of Home Inspectors (ASHI). Mr. Becker is the author of the best seller *The Complete Book of Home Inspection*, published by McGraw-Hill, and also writes the widely read "Homeowners Clinic" column for *Popular Mechanics* magazine. He has been qualified in court as an expert witness on the subject of home inspection, and appeared on national television on "Good Morning America" as a home inspection expert. Mr. Becker holds bachelor's and master's degrees in mechanical engineering and is a licensed Professional Engineer in New York, New Jersey, and Florida.